950

FLAT GLASS
TECHNOLOGY

FLAT GLASS TECHNOLOGY

RUNE PERSSON

LONDON

BUTTERWORTHS

ENGLAND: BUTTERWORTH & CO. (PUBLISHERS) LTD.
                LONDON: 88 Kingsway, W.C.2

AUSTRALIA: BUTTERWORTH & CO. (AUSTRALIA) LTD.
                SYDNEY: 20 Loftus Street
                MELBOURNE: 343 Little Collins Street
                BRISBANE: 240 Queen Street

CANADA: BUTTERWORTH & CO. (CANADA) LTD.
                TORONTO: 14 Curity Avenue, 16

NEW ZEALAND: BUTTERWORTH & CO. (NEW ZEALAND) LTD.
                WELLINGTON: 49/51 Ballance Street
                AUCKLAND: 35 High Street

SOUTH AFRICA: BUTTERWORTH & CO. (SOUTH AFRICA) LTD.
                DURBAN: 33/35 Beach Grove

Suggested U.D.C. number: 666·15

Suggested additional number: 691·615

Printed in Great Britain by Page Bros. (Norwich) Ltd.

PREFACE

GLASS is one of our oldest materials. In spite of this it is a most useful material in our modern world and one might well predict that its use will increase tremendously in the future. Through modern research work it has been possible to alter the properties of glass and to make it a most versatile material.

It is the purpose of this book to describe and define the different types of flat glass used in the building industry. The book is intended for people in the glazing business, architects, builders, glass technologists and all engineers working with glass products.

Besides being a reference book for professional people it can be used as a text-book for students in technical schools and universities.

Only Swedish standard specifications were included in the original Swedish edition. This edition includes British and American standard specifications. In general, this edition has been altered so as to make it more useful for the international reader.

Data included in this book have been contributed by many glass companies. The author is especially grateful to his own company, The Grängesberg Co., Stockholm, and to Pilkington Brothers Ltd., St. Helens, Lancashire, England and Pittsburgh Plate Glass Company, Pittsburgh, Pennsylvania, USA.

v

CONTENTS

CONTENTS

CONTENTS

1

HISTORY

A VOLCANIC glass called obsidian was created by nature a long time before man learned how to make glass. This natural glass was used to make arrow heads, knives and similar tools. Although it is not known when glass was first made artificially, it can be estimated that glass was made in Persia 7000 years ago. Fragments have been found mostly in Egypt, however, and it is known that glass was manufactured in that country around 2000 B.C. It appears that glass was first used as gems and later made into hollow vessels such as jars and vases.

The first use of glass in architecture probably occurred during the time of the Roman Empire. Glass mosaic was used for the decoration of walls and glass was also used in the windows of bath houses. One of the oldest glass windows that has been found was used in a bath house in Pompeii. It is a circular glass sheet with a diameter of about 13 cm (5 in) and is mounted in a bronze frame. The glass was made by a casting process and then drawn with pincers. Window openings up to a size of 200 × 200 cm (80 × 80 in) were planned for a public bath house being built just before Pompeii was destroyed in the eruption of Vesuvius in 79 A.D.

In the first centuries and up to about 1000 A.D., flat glass was mainly used for windows in churches and convents. Waxed paper, animal bladders and ground mica and alabaster were widely used in window openings of private houses in Europe even in the nineteenth century. The method of spinning window crowns (*Figure 1.1*) was probably first discovered in Syria in the eighth century. The glass workers blowpipe, used for this process, was invented at the beginning of the Christian era.

The well-known Venetian glass industry dates back to the tenth century. The special glass-making technique of Venice was kept secret, for a long period of time, by the guildsmen of Murano. The glass industry was moved to the island of Murano at the end of the thirteenth century in order to keep it secret. The glass makers of Murano did not specialize in flat glass products but a certain amount of window glass was manufactured and part of it was exported. In the fourteenth century mirrors were made by coating plates of glass with an amalgam of tin and mercury.

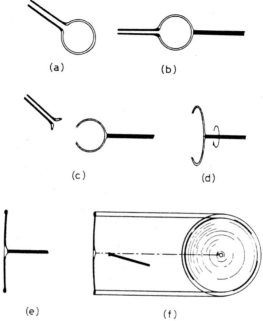

Figure 1.1. Manufacture of sheet glass according to the crown method; (a) at first a glass sphere or bubble is blown at the end of a blow-pipe; (b) a punty iron is attached to the bubble opposite the blow-pipe; (c) the blow-pipe is cracked off; (d) by heating and rotating the bubble will open up; (e) a flat circular sheet of glass is obtained; (f) the punty is removed; the circular sheet (the crown) has a thick edge and a thick central part (the bullion, or bull's eye); from R. Seiz, Glaser-Fachbuch (1963) by courtesy of Verlag Karl Hofmann, Schorndorf bei Stuttgart

The art of glass making was introduced into England at the beginning of the thirteenth century, probably by French glass makers. Glass making then successively spread to other parts of Europe and reached northern countries, like Sweden, in the sixteenth century. It then flourished in all parts of Europe.

English settlers introduced glass making into America. In fact, the first manufacturing establishment in America was a glass factory. This was erected at the beginning of the seventeenth century at James Towne, Virginia. The crown method of manufacturing flat glass was successively replaced by the cylinder process (*Figure 1.2*).

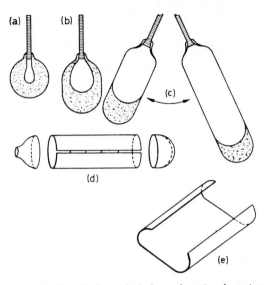

Figure 1.2. The cylinder method of manufacturing sheet glass; (a) a sphere is blown at the end of a blow pipe; (b) and (c) the sphere is elongated to a cylinder by blowing and swinging; (d) both ends are cut off and the cylinder is split; (e) the reheated cylinder is opened up; from H. Jebsen-Marwedel, Tafelglas (1950) *by courtesy of* Verlag Girardet, Essen

Larger sheets of glass could be made in this way and it was the dominating method of making flat glass in the nineteenth century. At the beginning of the twentieth century the machine-cylinder method was introduced. A circular metal bait about 25 cm (10 in) in diameter situated at the end of a blow-pipe, was lowered to the surface of the molten glass. Using compressed air for blowing, it was possible to draw a cylinder of glass, approximately 1500 cm (50 ft) high. This cylinder was subsequently split, flattened and annealed.

Many attempts were made at the end of the nineteenth century to draw a flat sheet of glass directly, so as to avoid the second process of flattening the cylinder. Such a process would simplify the flat glass

3

manufacturing technique and in addition should give a glass sheet of improved optical quality. The first workable process was invented by Emile Fourcault in Belgium. He took out his first patent in 1904 and ten years later the process had been fully developed to a manufacturing state. At approximately the same time, two methods for sheet glass drawing were developed in America. These were the Colburn, or Libbey-Owens, process and the Pittsburgh process. These processes are still used for making sheet glass, the most successful being the Pittsburgh process. For shop windows, mirrors, etc., a glass with more superior qualities than can be obtained by any of the sheet glass processes is often required. In order to obtain undistorted vision it is necessary that the two surfaces of the sheet are perfectly flat and parallel. These conditions are obtained by grinding and polishing ordinary sheet glass.

Special processes have, however, been developed for the manufacture of ground and polished plate glass. A casting process was invented in France in the middle of the seventeenth century. The glass was melted in pots and poured onto a casting table. The table was covered with fine sand to prevent sticking of the glass. By means of a heavy roller the glass was flattened to the required thickness. The glass was then ground, first with coarse sand and water and later with finer sand. The glass was then polished by felt-shod wheels fed with a fine abrasive rouge (iron oxide). About half the original thickness of the cast glass was removed by grinding and polishing.

The semi-continuous Bicheroux method was developed in Germany after World War I. By this method the glass was melted in pots and rolled not on a table but between two rollers. Continuous operations of grinding and polishing have been developed by the Ford Motor Company of America and by Pilkington Brothers Limited of England. In both these processes the glass has to be cut at the end of the annealing lehr and laid in separate pieces on the table of the continuous grinder and polisher. In 1938 Pilkingtons developed the twin grinding and polishing process. The glass was rolled from the tank in a continuous ribbon. After having been annealed in the lehr the glass ribbon passed through a twin grinder, where both sides were ground simultaneously, and then through a twin polisher. In 1959 Pilkingtons introduced their revolutionary float glass method (see Chapter 2). Glass made by this process, called float glass, combines the high surface finish of sheet glass with the flatness and lack of distortion of plate glass.

In recent years research has resulted in the creation of new improved

Figure 1.3. A disk of crown glass; from R. McGrath *and* A. C. Frost *by courtesy of* Architectural Press

types of flat glass product. Chemically-strengthened, photosensitive and crystallized glass are new types that will greatly increase the use of flat glass.

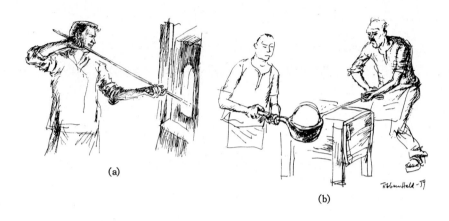

(a)

(b)

(d)

(c)

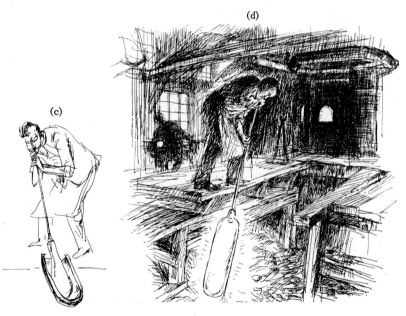

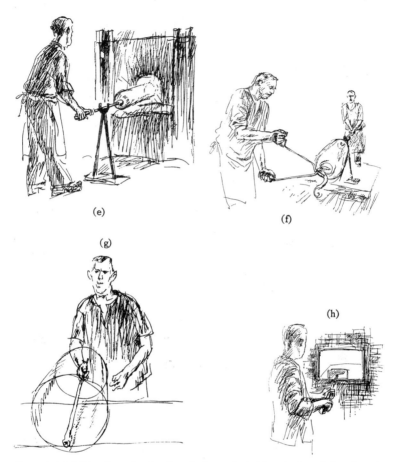

Figure 1.4. Making sheet glass by the cylinder method; (a) *gathering of glass from the pot in the furnace;* (b) *shaping the sphere in a wooden mould;* (c) *blowing in a mould to obtain a cylinder;* (d) *blowing and swinging to get the correct size of the cylinder;* (e) *reheating of the cylinder;* (f) *the cylinder ends are cut off by a pair of scissors;* (g) *the cylinder is split lengthwise;* (h) *the cylinder is reheated and flattened out to a sheet; by courtesy of* Oxelösunds Järnverk

BIBLIOGRAPHY

McGRATH, R. and FROST, A. C. *Glass in Architecture and Decoration,* Architectural Press, London, 1961
TOLEY, F. V. *Handbook of Glass Manufacture,* Vols. 1 and 2, Ogden, New York, 1960–1961

7

2

MANUFACTURE

2.1 GENERAL

THE four main types of flat glass product are sheet, plate, float and cast glass. The principal steps in the manufacture of these glasses are shown below.

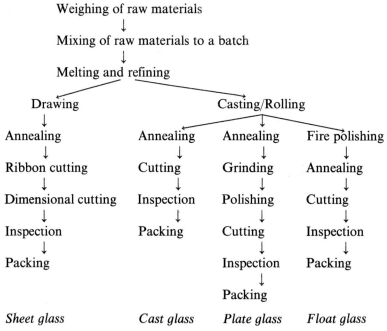

Weighing of raw materials
↓
Mixing of raw materials to a batch
↓
Melting and refining

Drawing   Casting/Rolling

Annealing   Annealing   Annealing   Fire polishing
↓   ↓   ↓   ↓
Ribbon cutting Cutting   Grinding   Annealing
↓   ↓   ↓   ↓
Dimensional cutting Inspection Polishing Cutting
↓   ↓   ↓   ↓
Inspection   Packing   Cutting   Inspection
↓   ↓   ↓
Packing   Inspection   Packing
↓
Packing

Sheet glass   *Cast glass*   *Plate glass*   *Float glass*

2.2 RAW MATERIALS AND MELTING

The raw materials required for the manufacture of glass are usually stored in silos, from which they are transported into the batch house. They are then weighed accurately and mixed to a batch. The following is a typical batch composition for a flat glass.

8

	kg (lb)
Sand	1 000
Soda ash	310
Limestone	55
Dolomite	240
Feldspar	60
Sodium sulphate	15
Cullet	400

The raw materials introduce different oxides into the glass and they may be classified in three categories according to the part played in forming the glass structure: Glass formers, Stabilizers, Fluxes. Different agents may also be added to the batch, for finishing and colouring purposes.

Sand consists of almost 100 per cent silica, which is the actual glass former. The stabilizing oxides calcium oxide, magnesium oxide and alumina, are introduced by limestone, dolomite and feldspar, respectively. Soda ash is a fluxing agent which decomposes to sodium oxide. Sodium sulphate is an agent which helps in refining and homogenizing the glass during the melting process.

The batch is fed into the glass melting furnace (*Figure 2.1*). A tank furnace for melting flat glass usually contains 500–2000 tons of glass and is made up of refractory materials of different types and qualities. The temperature in the melting end of a furnace is 1500–1550°C (2720–2820°F) and the furnace is usually heated with fuel oil or natural gas. Electric melting is also used and, in this case, the heat is generated in the glass by an electric current passing between electrodes situated in the glass bath. Some of the raw materials decompose during melting and give off large quantities of gases. When these gases pass up through the molten glass they help in stirring and homogenizing the glass. It is very important that the final glass has a homogeneous composition, in order to obtain constant physical properties. Before the glass is removed from the furnace it must also be free from bubbles, seeds and unmelted particles.

2.3 THE FOURCAULT METHOD

In the Fourcault method (*Figure 2.2*) the glass sheet is drawn vertically from a slot in a refractory floater or debiteuse, which floats on the glass in the drawing chamber at the end of the glass melting furnace. The debiteuse is forced down to a prescribed level in the glass and due to

B

*Figure 2.1. Feeding the batch into a continuous tank furnace for melting plate glass;
by courtesy of* Pittsburgh Plate Glass Company

the hydrostatic pressure developed in the slot the glass wells up through it. The drawing operation is started by means of a bait consisting of a metal frame, which is lowered into the slot. Since the glass adheres to the bait a continuous ribbon of flat glass can be drawn upwards. To maintain a uniform width of the glass ribbon, adjustable water-cooled metal boxes are placed on either side of it. The sheet of glass

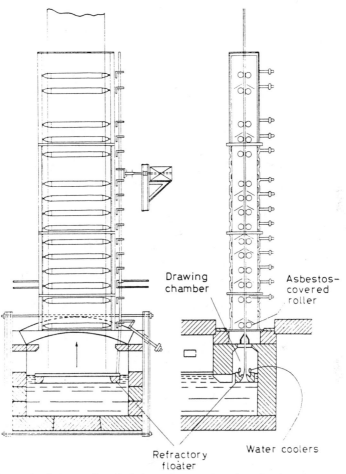

Drawing
chamber

Asbestos-
covered
roller

Refractory
floater

Water coolers

Figure 2.2. The Fourcault method of manufacturing sheet glass; from R. McGrath
and A. C. Frost *by courtesy of* Architectural Press

is drawn continuously through the drawing machine by means of a
series of asbestos-covered steel rollers stationed above the debiteuse.
After the glass ribbon has passed up through the annealing tower of
the machine it is cut into sheets of a predetermined length (lehr end
sizes). When starting the drawing the glass has a temperature of
about 950°C (1740°F) and at the top of the drawing machine its
temperature is 100–150°C (210–300°F). Most drawing machines

11

are approximately 7 m (23 ft) high. Cutting of the sheets to customer sizes may be carried out by automatic cutting machines or by hand (*Figure 2.3*). The edges of the glass ribbon are thicker than the actual sheet and these bulbs must therefore be removed. The machine widths usually vary between 140 cm and 280 cm ($4\frac{1}{2}$–9 ft) and the thickness of the glass from 1·5 mm to 7 mm ($\frac{1}{16}$–$\frac{9}{32}$ in).

Figure 2.3. Manual removal of glass panes at the cutoff floor of a Fourcault machine; by courtesy of Oxelösunds Järnverk

The speed of drawing 3 mm ($\frac{1}{8}$ in) thick glass is approximately 60–90 m/h (195–295 ft/h). The thickness of the glass is, in fact, controlled by the speed of drawing and increases as the speed decreases. Due to devitrification the drawing period for a Fourcault machine is limited to approximately 1 week.

2.4 THE LIBBEY-OWENS METHOD

The Libbey-Owens method (*Figure 2.4*) for drawing sheet glass was developed by Colburn in America. No refractory floater is used in

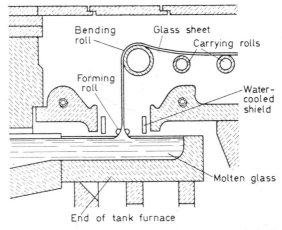

Figure 2.4. The Colburn (or Libbey-Owens) method of manufacturing sheet glass; by courtesy of Libbey-Owens-Ford Glass Company

this method to draw the sheet of glass. The width of the sheet is maintained by means of rollers which grip its edges. The sheet is pulled straight upwards from the glass surface in the drawing chamber of the tank furnace. After a short distance the glass ribbon is bent over a bending roll to a horizontal position. It then passes into a horizontal annealing lehr.

The machines usually have a width of 270–360 cm ($8\frac{1}{2}$–12 ft) and can draw sheets with a thickness of 1–30 mm ($\frac{1}{32}$–$1\frac{1}{4}$ in) and greater. The speed of drawing decreases with the thickness of the sheet and is approximately 120 m/h (395 ft/h) for a 3 mm ($\frac{1}{8}$ in) thick sheet.

13

2.5 THE PITTSBURGH METHOD

The Pittsburgh method (*Figure 2.5*) has been developed in America by the Pittsburgh Plate Glass Company. The sheet of glass is drawn vertically, directly from the surface of the molten glass in the drawing chamber of the tank furnace. To assist in conditioning the glass a submerged refractory draw bar is placed in the glass just below the

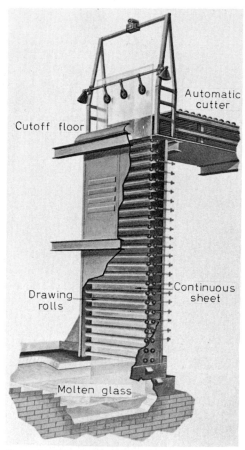

Figure 2.5. The Pittsburgh process for manufacturing sheet glass; a draw bar is submerged in the glass below the point of drawing; by courtesy of Pittsburgh Plate Glass Company

14

line of drawing. Air-cooled knurled rollers grip the edges of the sheet so as to maintain its width.

The Pittsburgh process is regarded as the most successful method for drawing sheet glass. The speed of drawing is high and it is possible to maintain a high quality during a drawing period of several weeks.

The width of the sheet drawn in a Pittsburgh machine is usually 220–320 cm ($7\frac{1}{4}$–$10\frac{1}{2}$ ft).

One modern version of this process is the Pennvernon method developed by the Pittsburgh Plate Glass Company. The speed of drawing 3 mm ($\frac{1}{8}$ in) thick glass is 70–110 m/h (235–365 ft/h). The thickness (d) varies with the speed of drawing (v) approximately according to the equation $v \cdot d^n = k$, where n and k are constants.

2.6 CAST GLASS

In the early method of making cast glass (*Figure 2.6*), the glass was melted in a pot and then poured out onto a casting table. The glass was then spread into plates by heavy rollers.

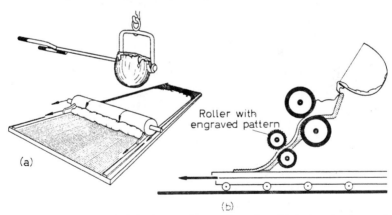

Roller with engraved pattern

(a)

(b)

Figure 2.6. Earlier methods for the manufacture of cast glass; (a) the hot glass is poured from the pot onto a casting table and then flattened by a movable roller; (b) the glass passes between two pairs of rollers onto a moving table; the second top roller is engraved so that a figured, rolled glass will be produced; after R. Seiz, Glaser-Fachbuch (1913) by courtesy of Verlag Karl Hofmann, Schorndorf bei Stuttgart

Later developments were the Bicheroux method and the P.P.G. ring-roll process. In these methods the glass was melted in pots and cast between rollers.

Continuous methods (*Figure 2.7*) have now been developed. The glass is melted in tank furnaces and drawn between rollers and it passes in a horizontal position through an annealing lehr. The gap between the water-cooled rollers is adjustable and usual thicknesses are between 4 mm and 20 mm ($\frac{5}{32}$–$\frac{13}{16}$ in). Ribbons of glass up to 350 cm (11$\frac{1}{2}$ ft) wide can be produced at a speed of drawing exceeding 360 m/h (1 180 ft/h).

Figure 2.7. A continuous process for manufacturing rolled glass; by courtesy of Pilkington Brothers Limited

The ordinary continuous or intermittent rolling processes give a rough-cast glass (or rough-rolled glass). If one of the rollers has a pattern, a figured rolled glass will be produced. By inserting a wire mesh during the process of manufacture a wired glass will be obtained (*Figure 2.8*). The wire mesh is drawn in on top of a ribbon of glass and

16

after passing a second pair of rollers, another ribbon of glass is formed on top of the wire. The wire is thus centrally embedded in the final product.

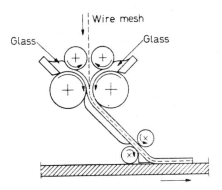

Figure 2.8. The principle of manufacturing wired glass; the wire is embedded between two layers of glass; after E. Steenberg and B. Simmingsköld *by courtesy of* Naturoch Kultur

2.7 PLATE GLASS

There are certain defects present in most sheet glass. A small variation in thickness is almost unavoidable, and this variation may cause distortion of vision. The two sides of a sheet must be absolutely flat and parallel in order to get perfect undistorted vision through a glass. This condition may be obtained after grinding and polishing cast glass. The first part in the manufacture of plate is the production of rough cast glass. The second parts are the grinding and polishing operations (*Figure 2.9*).

Until about 1920 plate glass was produced by the method of pot casting followed by grinding and polishing of one side at a time.

In 1938 Pilkington Brothers Limited succeeded in manufacturing plate glass in one continuous operation. The glass is melted in a tank furnace and rolled to a rough cast ribbon. This ribbon passes through an annealing lehr to remove all strain in the glass. The two surfaces are then ground simultaneously using sand and emery suspensions ranging from coarse to very fine grains. The twin grinders extend the full width of the ribbon. After leaving the twin grinders the ribbon

17

continues uncut through the twin polishing-machine units. Felt pads and a suspension of rouge in water are used for polishing the glass.

A twin grinder and polisher plant is more than 300 m (1000 ft) long.

Figure 2.9. Grinding and polishing of plate glass; by courtesy of Pilkington Brothers Limited

2.8 FLOAT GLASS

In 1959 Pilkington Brothers Limited presented their float glass process (*Figure 2.10*). This process makes it possible to produce a ribbon of glass having perfectly flat and parallel surfaces without grinding and polishing.

The glass is melted in a tank furnace and a ribbon is formed by passing the glass through a pair of rollers. This first part of the process is similar to the continuous cast glass process. The ribbon of glass is then drawn across the surface of molten tin. Heat is applied to both surfaces of the ribbon and this 'fire polishing' gives perfectly flat and parallel surfaces. An inert gas atmosphere protects the tin from oxidation and, therefore, it does not adhere to the lower surface of the ribbon. The glass ribbon is then annealed and cut in the ordinary way.

18

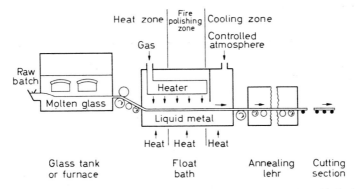

Figure 2.10. The float glass process; by courtesy of Pilkington Brothers Limited

Float glass combines the high surface finish of sheet glass with the flatness and lack of distortion of ground and polished plate glass. The float glass process is gradually replacing part of the plate glass processes.

The 'natural' thickness when making float glass is 6 mm ($\frac{1}{4}$ in) but glass of other thicknesses can also be made. The speed of drawing is high; a 3 mm ($\frac{1}{8}$ in) thick glass can be drawn at a speed of about 500 m/h (1 650 ft/h). A float glass line may run for 5 years turning out a 2 500 km (1 500 miles) ribbon of glass 3·3 m (11 ft) wide each year. Float glass is now produced in many different countries. Surface treatments, e.g. the introduction of coloured metal ions, can be carried out in the float bath.

BIBLIOGRAPHY

EITEL, W. and PIRANI, M. *Glastechnische Tabellen,* Springer, Berlin, 1932

GIEGERICH, W. and TRIER, W. *Glasmaschinen,* Springer, Berlin, 1964

JEBSEN-MARWEDEL, H. *Tafelglas,* Verlag Girardet, Essen, 1950

MCGRATH, R. and FROST, A. C. *Glass in Architecture and Decoration,* Architectural Press, London, 1961

PHILIPS, C. J. *Glass, its Industrial Applications,* Reinhold, New York, 1961

PIGANIOL, P. *Les Industries Verrieres,* Dunod, Paris, 1966

SHAND, E. B. *Glass Engineering Handbook,* York, Pa., 1958

STEENBERG, E. and SIMMINGSKÖLD, B. 'Glas', *Natur Kult., Stockh.,* 1958

TOLEY, F. V. *Handbook of Glass Manufacture,* Vols. 1 and 2, Ogden, New York, 1960–1961

3
CHEMICAL PROPERTIES

3.1 THE COMPOSITION OF GLASS

PRACTICALLY all flat glass is a soda-lime-silica glass, with a chemical composition within the following limits

Silica (SiO_2)	71–73%
Alumina (Al_2O_3)	0·5–1·5%
Iron oxide (Fe_2O_3)	0·05–0·15%
Calcium oxide (CaO)	5–10%
Magnesium oxide (MgO)	2–5%
Sodium oxide (Na_2O)	13–16%
Potassium oxide (K_2O)	0–1%
Sulphur trioxide (SO_3)	0–0·5%

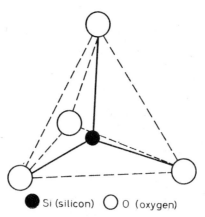

● Si (silicon) ○ O (oxygen)

Figure 3.1. The SiO_4 tetrahedron—the building brick of the glass structure

The basic units of the glass structure are the tetrahedral groups of SiO_4 in which the central silicon atom is surrounded by four oxygen atoms (*Figure 3.1*). The SiO_4 tetrahedra form a three-dimensional network in the glass. In crystalline forms of silica, e.g. quartz, the SiO_4 tetrahedra form a regular network. In glass, however, the network is irregular. This indicates that glass is not a crystalline substance (*Figure 3.2*).

In a physical sense glass is not a solid, but rather a supercooled liquid. Glass has therefore no melting point. According to the American Society for Testing Materials (ASTM) glass can be defined

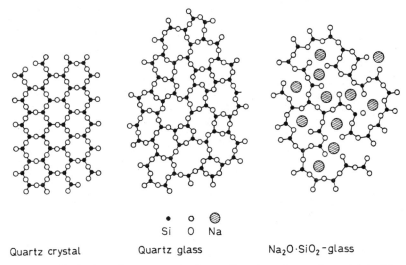

Si   O   Na

Quartz crystal   Quartz glass $Na_2O \cdot SiO_2$ -glass

Figure 3.2. A schematic representation in two dimensions of a quartz crystal, quartz glass and a $Na_2O–SiO_2$ glass

as an inorganic substance of fusion which has cooled to a rigid condition without crystallizing. At room temperature, however, glass has such a high degree of viscosity that it can be regarded as solid for all practical purposes. The rate of change of viscosity with temperature is shown in *Figure 3.3*.

The temperature ranges of some typical viscosity reference points are as follows:

Working range	650–1 100°C	(1 200–2 010°F)
Softening point	675–725°C	(1 250–1 340°F)
Deformation temperature	550–600°C	(1 020–1 110°F)
Annealing point	500–550°C	(930–1 020°F)
Strain point	450–500°C	(840–930°F)

21

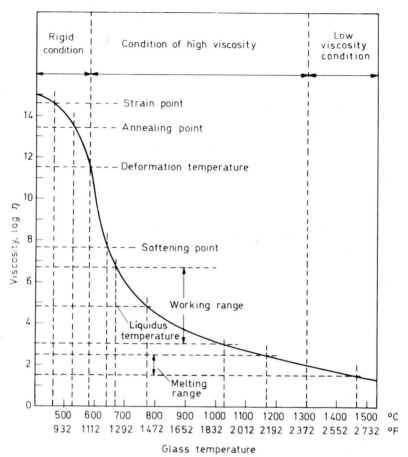

Figure 3.3. A typical viscosity curve for glass; from E. Steenberg and B. Simmings-köld[13] *by courtesy of* Natur och Kultur

The temperatures given above are only approximate since there are different definitions of the various reference points.

3.2 CHEMICAL DURABILITY

Glass, in general, has a very high resistance to the corroding action of water and atmospheric agencies. Glass windows will therefore withstand the action of atmospheric agencies for many years. Under certain conditions, however, glass will be corroded and it is not correct to speak of glass as a completely insoluble material. Glass is readily dissolved by hydrofluoric acid, which is used in the etching of glass. Its chemical resistance towards the attack of most other acids is very good, but alkaline solutions, on the other hand, attack glass quite severely.

By decreasing the alkali content of a glass its chemical durability is improved. Although a window glass of a soda-lime-silica type will

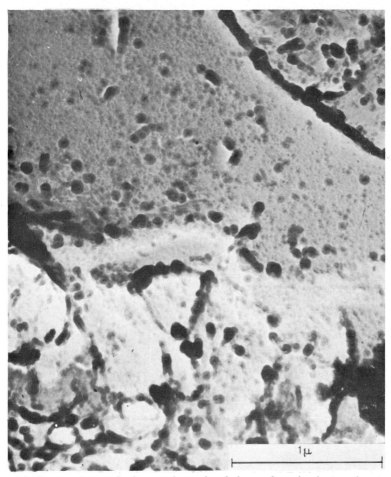

Figure 3.4. Electron micrograph of a severely weathered glass surface; the glass was kept in a storage room with a moist atmosphere for several months; by courtesy of Swedish Institute for Silicate Research

withstand the corroding action of rain for years it may be heavily corroded by a comparatively small amount of water retained for a short period in contact with the glass surface (*Figure 3.4*). In closely-packed window panes water may enter between the panes by capillary action. The water attacks the glass surface slightly and this produces a solution of sodium hydroxide, since the first compound to be sucked out by the water is sodium oxide. Ordinary rain could attack glass in the same way. Since rain water quickly disappears from the glass surface, however, the sodium hydroxide formed will have no chance of attacking the surface. If water is accumulated between closely-packed panes of glass, however, it may remain there for a considerable time. The alkaline solution of sodium hydroxide will corrode the glass heavily and give rise to a permanent staining. A film of sodium carbonate is formed when the sodium hydroxide reacts with carbon dioxide from the air.

If the attack has not proceeded too far it may be possible to remove the deposit on the glass surface by washing with water or dilute acid. It is possible to improve storage conditions by placing a sheet of paper of a suitable quality between the panes.

Since chemical attack on glass by water or an alkaline solution will remove part of the sodium oxide from the glass surface, this will improve the chemical durability of the surface[1]. It is known that attack by fumes of sulphur oxides in the drawing machines improves the durability of the surface. Sulphur trioxide can react with sodium oxide on the surface to give a deposit of sodium sulphate. This 'bloom' of sodium sulphate can easily be washed off leaving a clear surface with improved chemical durability. If the attack is too heavy, however, permanent staining may result.

In some cases an attacked glass surface may be iridescent. If the attack on the glass surface is not too deep it is possible to obtain a clear surface by the use of a polishing powder.

3.3 ATTACK BY CEMENTITIOUS WATER

Lime and cementitious water will corrode glass very heavily. Rain water after having been in contact with a new cement joint or concrete may become very corrosive and it is important, therefore, to prevent it from flowing over the glass windows. It is usually very difficult to remove a cement deposit from a glass surface. A glass surface which has been attacked by cementitious water is shown in *Figure 3.5*.

Figure 3.5. Electron micrograph of a glass surface attacked by a cementitious solution; by courtesy of Swedish Institute for Silicate Research

REFERENCES AND BIBLIOGRAPHY

[1] PERSSON, R. 'Improvement of the General Durability of Soda-lime-silica Glass Bottles by Treating with Various Agents', *Glass Technol., Sheffield,* **3** (1962)

[2] EITEL, W. and PIRANI, M. *Glastechnische Tabellen,* Springer, Berlin, 1932

[3] TOLEY, F. V. *Handbook of Glass Manufacture*, Vols. 1 and 2, Ogden, New York, 1960–1961

[4] PHILIPS, C. J. *Glass, its Industrial Applications*, Reinhold, New York, 1960

[5] SHAND, E. B. *Glass Engineering Handbook*, Maple Press, York, Pa., 1958

25

C

[6] McGRATH, R. and FROST, A. C. *Glass in Architecture and Decoration,* Architectural Press, London, 1961

[7] VÖLCKERS, O. *Tafelglasdaten,* Verlag Karl Hofmann, Schorndorf bei Stuttgart, 1954

[8] SEIZ, R. *Glaser Fachbuch,* Verlag Karl Hofmann, Schorndorf bei Stuttgart, 1963

[9] SPICKERMANN, H. *Erweitertes Gussglas Tabellarium*, Verlag Karl Hofmann, Schorndorf bei Stuttgart, 1958

[10] JEBSEN-MARWEDEL, H. *Tafelglass,* Verlag Girardet, Essen, 1950

[11] MOREY, G. W. *The Properties of Glass,* Reinhold, New York, 1945

[12] PIGANIOL, P. *Les Industries Verrieres,* Dunod, Paris, 1966

[13] STEENBERG, E. and SIMMINGSKOLD, B. 'Glas', *Natur Kult., Stockh.,* 1958

4

MECHANICAL PROPERTIES

4.1 INTRODUCTION

THE mechanical strength of a glass product depends to a great extent on the condition of its surface. Chemical or mechanical attack on a glass surface immediately reduces the strength of the product. It is very difficult, therefore, to give an exact value of the mechanical strength. Theoretically, the tensile strength of flat glass is approximately 1000 kgf/mm^2. In practice, however, it has been found that flat glass products have a strength equal to 1% of the theoretical value. It has been proved that microscopic or submicroscopic surface scratches or flaws reduce the strength of glass considerably. Tests with glass fibres have shown that the tensile strength can be increased by etching the fibre using hydrofluoric acid. This gives a virgin surface free from scratches and flaws. The strength of the fibre may be increased up to about 50% of the theoretical value. Unless the fibres are kept in a vacuum their strength will decrease rapidly. Much research work is being carried out to try to improve the mechanical properties of glass. It can, therefore, be expected that the strength of glass products will be increased considerably in the future.

The compressive strength of glass is approximately 10 times its tensile strength. Due to the brittle character of glass, however, fracture is almost always caused by tension.

Due to its brittle condition there is no plastic deformation in glass which will break suddenly when subjected to a stress exceeding its elastic limit. Glass obeys Hooke's law accurately until the stress is great enough to cause fracture.

4.2 WEIGHT

The density of flat glass is 2·5 kg/dm^3 (155 lb/ft^3). For most practical calculations it can be said that the specific gravity of glass is 2·5. It is thus a comparatively light material, lighter than aluminium, the specific gravity of which is 2·7. One square metre of a 2 mm thick sheet of glass weighs 5 kg. (One square foot of $\frac{1}{8}$ in thick sheet glass weighs 26 oz.) The weights per unit area of glass of different thicknesses are shown in *Table 4.1*.

Table 4.1. *Approximate weights per unit area of glass of different thicknesses*

Metric units		British units	
Thickness (mm)	Weight (kg/m²)	Thickness (in)	Weight (oz/ft²)
2	5·0	$\frac{1}{16}$	13
3	7·5	$\frac{3}{32}$	19
4	10·0	$\frac{1}{8}$	26
5	12·5	$\frac{5}{32}$	32
6	15·0	$\frac{3}{16}$	39
7	17·5	$\frac{1}{4}$	51
10	25·0		

Since the thickness of glass may vary somewhat within a certain range of tolerances it may be recommended to use a specific gravity of 2·6 for glass in constructional work.

4.3 THE INFLUENCE OF STRAIN ON THE STRENGTH OF GLASS

As mentioned in Section 4.1 the strength of glass will decrease considerably if surface scratches or flaws are present. Fractures always start at the glass surface and scratches will determine the point of origin and thus greatly reduce the apparent strength.

A scratch will affect the strength of a well-annealed glass more than that of a disannealed glass. The latter is not harder to scratch but it is more difficult to extend the scratch into a crack. The state of annealing is therefore of great importance to the strength of the glass. All ordinary flat glass products are well-annealed and are therefore almost completely free from strain otherwise it would be difficult to cut the glass panes. It is important to ensure that the surface of the glass is free from tension during manufacture.

The compressive strength of glass is approximately 10 times the tensile strength. Since fracture of glass is a result of its tensile strength having been exceeded it is important to prevent the development of permanent tensile strain on the surface of the glass. If the annealing process gives rise to permanent compression at the glass surface the mechanical properties of the product will be improved.

28

In order to develop a fracture at a glass surface which is under compression it is necessary first to neutralize the compression by tension and then to apply an additional tensile stress so as to exceed the tensile strength of the glass.

When a glass is cooled quickly, compression will develop at the surface with a counter-balancing tension in the interior of the glass. This is the principle used in tempering glass. After being heated to a temperature above its annealing point the glass is cooled rapidly. The strength of a tempered glass is four to five times that of annealed glass, and it cannot be broken unless sufficient force is applied to overcome the compression on the surface and, in addition, to exceed the tensile strength. When tempered glass breaks, many small interlocking fragments are formed.

Thermal and chemical strengthening of glass is described in Subsection 9.9.1.

4.4 HARDNESS

Glass is very hard in comparison with other transparent materials. It is much harder than plastics and harder than many metals, particularly under conditions of 'scratch-hardness' testing. Hardness may be defined in terms of scratchability as in Moh's scale of hardness. In this scale talc has hardness 1 and diamond 10, the highest figure of the scale and each material will scratch all those below it in the scale.

Table 4.2. Moh's scale of hardness

10	Diamond	5	Apatite
9	Corundum	4	Fluorite
8	Topaz	3	Calcite
7	Quartz	2	Gypsum
6	Feldspar	1	Talc

Glass has a hardness of 6 on Moh's scale. There may be a slight difference in hardness between different types of glasses, but they will all scratch each other. The hardness of toughened glass is equal to that of annealed glass.

4.5 THE MECHANICAL STRENGTH OF A PANE OF GLASS

The thickness of a window pane must be chosen with regard to its dimensions. It is also important to consider wind pressure on buildings. Glass manufacturers and national standard specifications give general recommendations on the thickness of a window pane in relation to its dimensions. A nomogram for determining the thicknesses

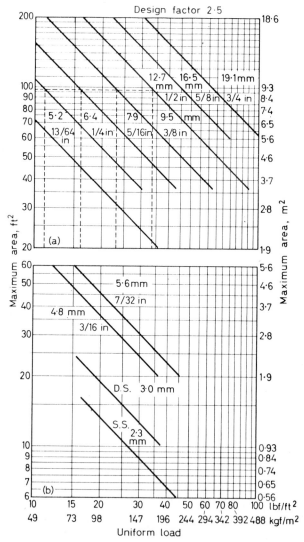

Figure 4.1. Design load data; (a) polished plate glass; (b) sheet glass; by courtesy of Libbey-Owens-Ford Glass Company

Graphs apply to square and rectangular lights of glass when the length is no more than five times the width. Values are based on L-O-F published minimum thicknesses. All edges continuously held.

Data shown are based on actual tests to destruction.

Example—Design conditions 97 ft^2 (9 m^2) glass area with 30 p.s.f. (147 kgf/m^2) code requirement. Dotted lines indicate 11 p.s.f. (54 kgf/m^2) capability for $\frac{1}{4}$ in (6·4 mm) thickness; 16 p.f.s. (78 kgf/m^2) for $\frac{5}{16}$ in (7·9 mm); 23 p.s.f. (112 kgf/m^2) for $\frac{3}{8}$ in (9·5 mm) and 34 p.s.f. (166 kgf/m^2) for $\frac{1}{2}$ in (12·7 mm). Therefore $\frac{1}{2}$ in (12·7 mm) thickness is recommended.

To assist the architect or engineer, the following are the predicted percentages of glass failure for various design factors. These percentages apply when a large number of lights are being considered.

Design factor	Breakage (%)
1·0	50·0
1·5	9·2
2·0	2·3
2·5	0·8
3·0	0·4
4·0	0·14
5·0	0·07
10·0	0·02

When a design factor other than 2·5 is chosen, these graphs may be used providing the design load is adjusted as follows:

$$\left(\begin{array}{c}\text{Design load for}\\ \text{use with graphs}\end{array}\right) = \left(\frac{\text{Actual design load}}{2·5}\right) \times \left(\begin{array}{c}\text{Chosen design}\\ \text{factor}\end{array}\right)$$

Example—Assume a building code requires the choice of glass areas and thicknesses to be based on a design load of 25 p.s.f. (122 kgf/m^2) and the architect or engineer decides the appropriate design factor is 2·0. The adjusted design load for use with the graphs would be determined as follows:

$$\text{Design load for use with graphs} = \left(\frac{25}{2·5}\right) \times (2·0) = 20 \text{ p.s.f. (98 kgf/m}^2)$$

of window panes exposed to various wind pressures is shown in *Figure 4.1.*

4.5.1 Calculation of thickness

It is difficult to give a general formula for the calculation of the thickness of a pane of glass. The maximum weight that a pane can withstand depends on several factors such as the ratio of the two sides, the glazing technique used, the loading time, etc. General formulae for the calculation of strength, may be used. The Marcus formula[1] is used in Subsection 4.5.3 in calculating the thickness of an aquarium wall. Another formula has been derived by Wigen[2], this formula states that

$$t = \sqrt{\left[\frac{0.75\,p\,a^2}{\sigma(1 + 1.6\,h^3)}\right]}$$

where $h = (a/b)$, a is the width of the pane in centimetres, b is the height of the pane in centimetres, p is the load in kilogrammes-force per square centimetre and σ is the strength of the glass in kilogrammes-force per square centimetre.

In most formulae for the calculation of the thickness of a glass pane only the total load and the ratio of the two sides are considered. The size of the pane, therefore, does not influence the result. The load is regarded as being evenly distributed on the surface of the pane. When choosing a value for the strength of the glass it is necessary to take account of the length of the time of loading, the safety factor, etc. Some strength values are given at the end of this chapter. The minimum thickness to be used for a window pane will always be determined with regard to its surface area (see *Figure 9.1*).

The required thickness of a window pane of dimensions 140 × 120 cm (55 × 47 in) which has to withstand a wind pressure of 110 kgf/m² (23 lb/ft²), can be found by using Wigen's formula ($\sigma = 300$ kgf/cm²) (4 250 lb/in²), and is given by

$$t = \sqrt{\left\{\frac{0.75 \times 0.011 \times 140^2}{300\left[1 + 1.6\left(\frac{140}{120}\right)^3\right]}\right\}}$$

therefore $\quad t = \sqrt{0.1515} = 0.39\text{ cm} \simeq 4\text{ mm }(\tfrac{5}{32}\text{ in})$

Table 4.3. *Permissible distributed loads for plate glass shelves*

Span (m) (ft)	0·3 1	0·6 2	0·9 3	1·2 4	1·5 5
Nominal thickness	Permissible load kgf/m² (lb/ft²)				
(mm) (in)					
4·7 $\frac{3}{16}$	203 (41·6)	43 (8·9)	14 (2·8)	3 (0·7)	—
6·4 $\frac{1}{4}$	403 (82·7)	90 (18·6)	33 (6·7)	12 (2·5)	3 (0·6)
9·5 $\frac{3}{8}$	1100 (226)	260 (53·1)	102 (21·0)	48 (9·8)	22 (4·6)
12·7 $\frac{1}{2}$	1882 (387)	450 (92·1)	184 (37·6)	90 (18·5)	47 (9·6)

From McGrath and Frost[1] by courtesy of *The Architectural Press*.

Approximate figures for the thickness of window panes as a function of the load they may bear can be found from *Figure 4.1*. The mechanical strength of glass has been assumed to be 150 kgf/cm² (2130 lb/in²).

McGrath and Frost[1] have calculated permissible uniform loads for various sizes of plate glass shelf. They used a value of 95 kgf/cm² (1340 lb/in²) for the modulus of rupture. Their results are shown in *Table 4.3*.

4.5.2 Wind velocity and wind pressure

The relationship between wind velocity and wind pressure is shown in *Table 4.4*.

Table 4.4

Wind velocity, v		Wind pressure, p	
(m/s)	m.p.h.	(kgf/m²)	(lb/ft²)
6	13·4	2	0·4
8	17·9	4	0·8
10	22·4	6	1·2
12	26·8	9	1·8
14	31·4	12	2·5
16	35·8	16	3·3
18	40·3	20	4·1
20	44·8	25	5·1
22	49·3	30	6·2
24	53·7	36	7·4
26	58·2	42	8·6
28	62·6	49	10·0
30	67·2	56	11·5

The wind pressure in kilogrammes per square metre is numerically equal to the wind pressure in millimetres of water gauge. The pressure at right angles to the wind direction can be calculated using the formula $p = v^2/16$.

4.5.3 Glass for an aquarium

Marcus[1] has suggested a formula for the calculation of the thickness of glass for an aquarium. The hydrostatic pressure is proportional to the depth of the water. It can be assumed, however, that there is a uniform pressure against the glass wall equal to the pressure at a quarter of the height of the wall. The Marcus formula can be written

$$f = \frac{3}{4} \cdot \frac{W}{t^2} \left[1 - \frac{5}{6} \cdot \frac{r^2}{1 + r^4} \right] \left[\frac{r(1 + \sigma r^2)}{1 + r^4} \right]$$

Table 4.5. Maximum glass sizes** of various thicknesses for glazing vertical windows supported on four sides [square feet (mm²) of area]; by courtesy of Pittsburgh Plate Glass Company

Glass thickness (in) (mm)	S.S.–0·087 2·3	DS–0·118 3·0	$\frac{1}{8}$ 3·2	$\frac{3}{16}$ 4·8	$\frac{1}{4}$ 6·4	$\frac{5}{16}$ 7·9	$\frac{3}{8}$ 9·5	$\frac{1}{2}$ 12·7	$\frac{5}{8}$ up to 1 15·9–25·4	$1\frac{1}{4}$ 31·8
30 Mile (13·5 m/s) Wind	35 ft²† 3·3 m²	64·5† 6·0	72† 6·7	162 15·1	198* 18·4	198* 18·4				
40 Mile (17·9 m/s) Wind	17·5† 1·6	32·25† 3·0	36† 3·3	81 7·5	144 13·4	198* 18·4	240* 22·4			
55 Mile (24·6 m/s) Wind	11·6 1·1	21·5† 2·0	24† 2·2	54 5·0	96 8·9	150 14·0	216 20·0			
65 Mile (29·0 m/s) Wind	8·7 0·8	16·1† 1·5	18† 1·7	41 3·8	72 6·7	112 10·4	162 15·0	240* 22·3		
80 Mile (35·7 m/s) Wind	5·8 0·5	10·85 1·0	12 1·1	27 2·5	48 4·5	75 7·0	108 10·0	192 17·9		
100 Mile (44·6 m/s) Wind	3·5 0·3	6·45 0·6	7 0·7	16 1·5	29 2·7	45 4·2	65 6·0	115 10·7		
120 Mile (53·5 m/s) Wind	2·5 0·2	4·6 0·4	5 0·5	11 1·0	20 1·9	32 3·0	46 4·3	82 7·6	80* 7·4	74* 6·9

† 'SS', 'DS' and $\frac{1}{8}$ in glass, because of flexibility, should never be used in areas exceeding 12 ft² (1·1 m²), and for the sake of appearance, areas used should not exceed 7 ft² (0·7 m²).

* Maximum size available.

** These sizes are based on practical experience and are entirely adequate. They should not be referred to the chart of safe loads for plate glass since in actual use, due to intermittent loading, the actual factor of safety is several times the apparent values.

where f is the modulus of rupture, in kilogrammes-force per square centimetre (or lb/in^2), W is the total load in kilogrammes or lb, carried by the pane, t is the thickness of the glass, in centimetres or inch, $r = b/a$, a is the longest side in centimetres or inches, b is the shortest side in centimetres or inches and σ is Poisson's ratio $= 0.25$.

The total load, W, can be calculated using the following formula

$$W = \frac{3}{4} \cdot h \cdot \frac{1}{1000} \cdot \rho \cdot a \cdot b$$

where h is the depth of water in the aquarium and ρ is the specific gravity ($= 1$ for water).

4.5.3.1. *Example*—The dimensions of a glass pane to be used in an aquarium are 100×50 cm (39×20 in) (50 cm is the height). What thickness can be recommended for the glass?

$$W = \frac{3}{4} \cdot 50 \cdot \frac{1}{1000} \cdot 100 \cdot 50 = 187.5 \text{ kg (414 lb)}$$

Using sheet glass, a reasonably safe value for the modulus of rupture is 100 kgf/cm^2 (1420 lb/in^2). Now

$$r = \frac{50}{100} = \frac{1}{2}$$

Therefore $\quad 100 = \dfrac{3}{4} \cdot \dfrac{187.5}{t^2} \left[1 - \dfrac{5}{6} \cdot \dfrac{(\frac{1}{2})^2}{1 + (\frac{1}{2})^4} \right] \left[\dfrac{\frac{1}{2}(1 + 0.25 \cdot (\frac{1}{2})^2)}{1 + (\frac{1}{2})^4} \right]$

giving $\quad t^2 = 1.406 \times 0.804 \times \frac{1}{2} = 0.565$

and $\quad t = 0.75$ (cm)

i.e. a glass thickness of 7.5 mm ($\frac{5}{16}$ in) is required.

4.6 GLASS FRACTURE

A fracture in glass is propagated at right angles to the tension producing it. Fractures have a very definite form and as a result it is possible to diagnose their cause. In general, it can be said that a high stress before rupture will produce a more complicated form of fracture than a lower stress. The speed of propagation through the glass is also dependent on the stress in the glass. A very low speed can be obtained by knocking carefully at the end of a fracture or by heating it. Fractures caused by severe impacts may have velocities up to 2000 m/s (2200 yard/s).

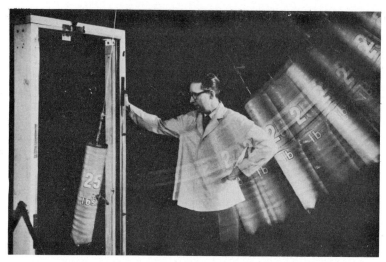

Figure 4.2. A free-swinging 25 lb dead weight fails to dent a safety glass in a glass patio door; by courtesy of Pittsburgh Plate Glass Company

A fracture surface shows two different types of mark, rib marks and hackle marks. The wave-fronts of rib marks point in the direction of propagation of the fracture (*Figure 4.3*). Hackle marks are formed at right angles to the direction of the fracture. These marks are only

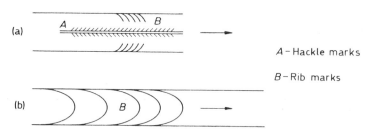

Figure 4.3. Fracture marks on glass; (a) high internal stress; (b) low internal stress

present when the fracture has been caused by high stress.

By studying the fracture surfaces it is thus possible to determine in which direction the fracture travelled and also the approximate stress in the glass before rupture occurred.

4.7 MECHANICAL PROPERTIES

Modulus of elasticity	$10–11.4 \times 10^6 \, lb/in^2$	$700,000–800,000 \, kgf/cm^2$
Tensile strength	$4250–14,200 \, lb/in^2$	$300–1000 \, kgf/cm^2$
Compressive strength	$130,000–140,000 \, lb/in^2$	$9000–10,000 \, kgf/cm^2$
Modulus of rupture	$4250–14,200 \, lb/in^2$	$300–1000 \, kgf/cm^2$
Impact, 4 mm ($\frac{5}{32}$ in) glass	8–45 in-lb	$0.1–0.5 \, kgf \, m$
Impact, 6 mm ($\frac{1}{4}$ in) glass	35–45 in-lb	$0.4–0.5 \, kgf \, m$
Poisson's ratio	0.25	0.25
Stress optical coefficient	2.5–2.7 Brewster	2.5–2.7 Brewster

Due to the large variability in the strength of glass it is necessary in solving practical problems to use strength values so low as to cover all but a negligible proportion of the weakest specimens. *Table 4.6* gives the maximum working stresses allowed for a risk of not more than 1 % for a glass thickness up to 4 mm ($\frac{5}{32}$ in).

Table 4.6

	Sustained loading		Momentary loading	
	(lb/in^2)	(kgf/cm^2)	(lb/in^2)	(kgf/cm^2)
Sheet glass	2 200	150	4 500	300
Float glass	1 500	110	3 000	210
Plate glass	1 400	100	2 500	175
Ornamental glass	1 100	75	2 100	150
Wired glass	1 000	70	2 000	140

REFERENCES AND BIBLIOGRAPHY

[1] McGRATH, R. and FROST, A. C. *Glass in Architecture and Decoration*, Architectural Press, London, 1961
[2] WIGEN, R. 'Winduer', *Handb. Norg. Byggforskningstinst.*, Oslo, 15 (1963)
[3] EITEL, W. and PIRANI, M. *Glastechnische Tabellen*, Springer, Berlin, 1932
[4] TOLEY, F. V. *Handbook of Glass Manufacture*, Vols. 1 and 2, Ogden, New York, 1960–61
[5] PHILIPS, C. J. *Glass, its Industrial Applications*, Reinhold, New York, 1960
[6] SHAND, E. B. *Glass Engineering Handbook*, Maple Press, York, Pa., 1958
[7] VÖLCKERS, O. *Tabelglasdaten*, Verlag Karl Hofmann, Schorndorf bei Stuttgart, 1954
[8] SEIZ, R. *Glaser-Fachbuch*, Verlag Karl Hofmann, Schorndorf bei Stuttgart, 1963

9 SPICKERMANN, H. *Erweitertes Gussglas Tabellarium,* Verlag Karl Hofmann, Schorndorf bei Stuttgart, 1958
10 JEBSEN-MARWEDEL, H. *Tafelglas,* Verlag Girardet, Essen, 1950
11 MOREY, G. W. *The Properties of Glass,* Reinhold, New York, 1945
12 PIGANIOL, P. *Les Industries Verrieres,* Dunod, Paris, 1966

5
OPTICAL PROPERTIES

5.1 COLOURLESS GLASS

ONLY part of a beam of light striking a glass wall will pass through it; some of the light is reflected at the front surface. The remainder of the light passes into the glass where part of it is absorbed and part reflected at the second surface. The percentage transmittance of light through a glass wall depends on the optical properties of the glass and on the wavelength of the incident light.

Almost all types of flat glass have a refractive index of approximately 1·5. The proportion of light that is reflected can be found by using the Fresnel formula

$$R = \left(\frac{n-1}{n+1}\right)^2$$

where R is the reflectance and n is the refractive index.

The reflectance will thus be 0·04 from each surface. With a reflectance factor of 0·08 or 8% the maximum transmittance of light through the glass is 92%. Since there is always some absorption of light by the glass the transmittance through all commercial glass products will be less than 92%. The absorptance varies with the wavelength of the light.

Transmittance curves for two sheet glasses are shown in *Figure 5.1*. It is seen from these curves that transmittance through a glass decreases with an increase in the iron content of the glass. The visible part of the light has a wavelength between 380 nm and 780 nm (1 nanometre $= 10^{-9}$ m $= 1$ millimicron, mμ). In this part of the spectrum 'commercially' colourless flat glass products have a light transmittance of 80–90%. As seen from *Figure 5.1* the transmittance is very much lower in the short wavelength region, i.e. the ultraviolet part of the spectrum. Light of wavelength shorter than 300 nm is not transmitted through ordinary soda-lime-silica glass. In the infra-red region (light of longer wavelength) the transmittance is of the same order as that in the visible region up to about 2000 nm.

39

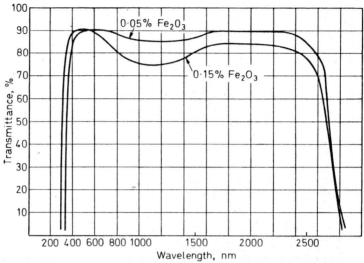

Figure 5.1. *The light transmittance at various wavelengths for two sheet glasses having different contents of iron oxide*

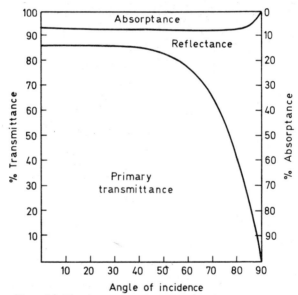

Figure 5.2. *The absorptance, transmittance and reflectance of a 3 mm thick sheet glass; from* G. B. Pleijel[13] *by courtesy of* Statens råd för Byggnadsforskning

The transmittance then decreases with increased wavelength. At wavelengths longer than 3000 nm there is a very small light transmittance through regular flat glass.

The transmittance, absorptance and reflectance vary with the angle of incidence of the light beam as seen from *Figure 5.2*. There is very little change, however, up to an angle of incidence of 40°.

In general, it can be said that the transmittance through sheet glass, plate glass and float glass is approximately 85–90% in the visible region of the spectrum. Rough cast glass transmits approximately 80% and ornamental glass 70–85%, depending on the pattern.

By special mechanical or chemical surface treatments it is possible to obtain glass giving reduced glare effects from lighting points. The surfaces of the glass are lightly textured and therefore give a 'diffuse' reflection. This type of glass is used for picture and photograph framing and for instrument panels (*Figure 5.3*).

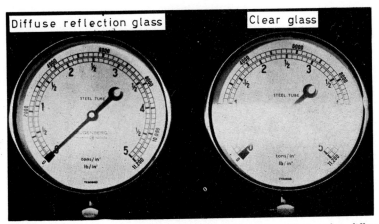

Figure 5.3. These dials, photographed under identical conditions, illustrate how diffuse reflection glass subdues reflected glare and allows clear vision for accurate dial reading; by courtesy of Pilkington Brothers Limited

5.2 LIGHT TRANSMITTANCE BY DIRTY SURFACES

All transmittance figures are related to clean windows. A window can only remain clean for a short period of time after cleaning. The transmission of light may decrease by as much as 10% before the windows are regarded as being 'dirty'. This is a factor that should be

41

borne in mind when calculating the quantity of daylight passing through any window. Approximate transmittance figures for dirty window panes are given in *Table 5.1*.

Table 5.1. *Light transmittance by dirty windows (approximate figures for double-glazed windows)*

Surface	Transmittance (%)
Clean surfaces	75–85
2 months after washing	70–80
Dirty 'residential' windows	65–75
Dirty 'industrial' windows	55–65
Very dirty windows—possible to look through	50–60
Very dirty windows—difficult to look through	30–40

5.3 COLOURED GLASS

Glass of many different colours can be made. The following colouring agents are commonly used.

Green—iron oxide, chromium oxide

Blue—cobalt oxide, copper oxide

Violet—manganese oxides

Amber—carbon and sulphur

Red—selenium and cadmium sulphide

Opal glass—fluorine compounds, e.g. cryolite

The metallic oxides are usually dissolved in the glass. These types of coloured glass are therefore transparent. Some other colouring agents like sulphur and selenium appear in glass as colloidal particles and may give translucent properties to the glass. The opal glasses usually contain crystalline particles, e.g. fluorides. These glasses owe their light scattering properties to the tiny crystalline inclusions which have different indices of refraction from that of the glass itself. Many coloured glasses are quite opaque to visible light.

In general, it can be said that colour is caused by absorption of various parts of the visible spectrum. A blue glass, for example, is blue because it transmits more readily the particular vibrations which bring the sensation of blue to the eye. The other parts of the light in the spectrum are absorbed or less readily transmitted. The intensity of the colour depends on the amount of colouring agent, e.g. cobalt oxide, present in the glass.

42

There are many different methods by which colours can be defined. Spectral transmittance curves give accurate definitions of coloured glasses. Such transmittance curves are shown in *Figure 5.1* and *Figure 5.4.*

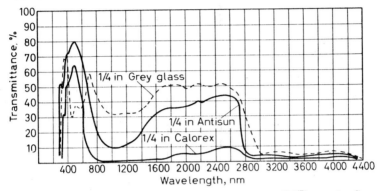

Figure 5.4. Different types of heat-absorbing glass; by courtesy of Pilkington Brothers Limited

The Munsell colour atlas gives a possibility of classifying colours by direct visual comparison. This system is used by the British Standards Institution (BSI). A colour can then be defined by a BSI number.

A widely used system of defining a colour is the Commission Internationale de l'Eclairage (CIE) system. In this system three imaginary colours x, y and z are used. Any colour can be defined by using the CIE properties denoting wavelength, brightness and purity.

By adding heavy metal oxides, such as lead and cerium oxides, to a glass, a high absorbtance of radioactive rays may be obtained. Such glasses may be used in radiation chambers and it is important therefore, that their colour is unchanged by irradiation.

5.4 THE INFLUENCE OF GLASS THICKNESS ON LIGHT TRANSMITTANCE

The transmittance of light through a glass decreases logarithmically with increased glass thickness. In a simplified form this function may be written as

43

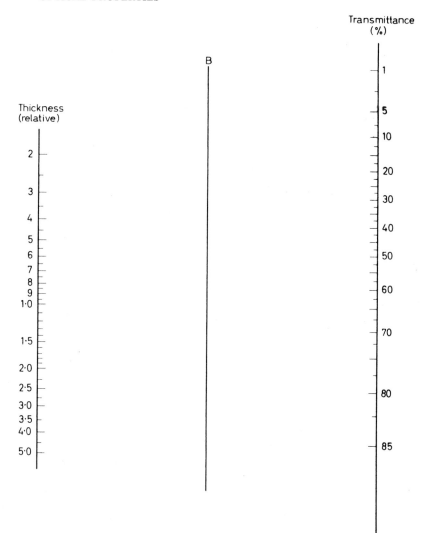

Figure 5.5. Nomogram for the determination of change of transmittance with change of thickness of glass; from E. B. Shand[6] by courtesy of McGraw-Hill Book Company

If a 2 mm thick coloured glass has a transmittance of 80% at a wavelength of 600 nm its transmittance for a thickness of 4 mm may be found by drawing a line from 1·0 relative thickness to 80% transmittance. From 2 (=4/2 mm) relative thickness draw a line, cutting line B at the same point as the first line drawn. The elongation of this second line will cut the line for transmittance at 68%. This, then, is the transmittance for the glass when its thickness is 4 mm.

$$\log \frac{T_2}{0 \cdot 92} = \frac{t_2}{t_1} \log \frac{T_1}{0 \cdot 92}$$

where T_1 is the transmittance of the glass when the thickness is t_1. Similarly T_2 is the transmittance of the glass of thickness t_2.

The nomogram in *Figure 5.5* can also be used to find the transmittance of any thickness of a glass, when the transmittance of one thickness is known.

5.5 HEAT-ABSORBING GLASS

As seen from *Figure 5.1*, addition of ferric oxide (Fe_2O_3) to a glass decreases its light transmittance. If the iron is present in a reduced form as ferrous oxide (FeO) its absorption of infra-red light is increased.

The iron oxides are present, to some extent, in all types of flat glass. By increasing the amount of iron oxides and by melting under reducing conditions so as to get a high proportion of the reduced form of the oxide (FeO), a heat-absorbing glass will be obtained. Different types of heat-absorbing glass can be manufactured by

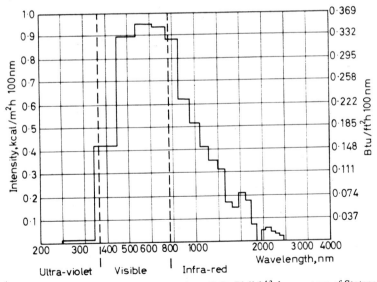

Figure 5.6. The intensity of solar rays; from G. B. Pleijel,[13] by courtesy of Statens råd för Byggnadsforskning

45

adding other metal oxides besides iron oxide to the glass batch. The oxides of nickel, cobalt and copper may be used.

Transmittance curves for some different heat-absorbing glasses are shown in *Figure 5.4.*

As seen from the transmittance curves there is a decrease of the transmittance of visible light as well. This is necessary in order to obtain good heat-absorbing qualities, since approximately 45% of the heat in the sun's rays comes from the visible part of the spectrum. The spectral transmittance curve for the sun's radiation is shown in *Figure 5.6.*

Heat-absorbing glasses may absorb 30–75% of the visible light and 50–90% of the infra-red light.

Technical data of some commercial heat-absorbing glasses are shown in *Table 9.6.*

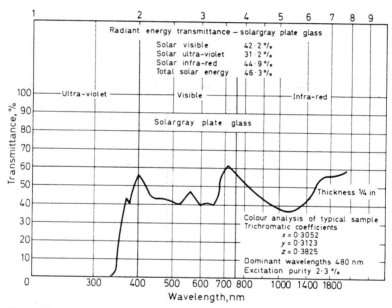

Figure 5.7. Transmittance curve for a heat-absorbing glass; by courtesy of Pittsburgh Plate Glass Company

5.6 LIGHT- AND HEAT-REFLECTING GLASSES

An untreated flat glass surface reflects about 4% of the light. By applying a metallic coating to the surface its light reflectance may

increase. Special light- and heat-reflecting glasses have been manufactured for about 25 years.

Electrically-conductive glasses were developed in the USA during
World War II to be used as sheets to protect personnel from intense
long-wave radiation. The ceramic coatings applied to these glasses
transmitted solar radiation but reflected long-wave, infra-red energy.
Similar ceramic coatings have been developed for window glass. The
transmittance for both visible and total solar radiation can be varied
between 70% and 15%[1].

Selective reflectance coatings can also be applied by vacuum
evaporation of metals onto the glass surface. Gold coatings are
usually applied in this way. The reflection of light from a gold coating
increases with increased wavelength. The transmittance of infra-red
light through a gold coating is therefore very small. On the other
hand, gold coatings give a characteristic gold colour to the glass, in
reflected light. In consequence, the transmitted light has a greenish
colour.

A third way of applying a selective coating to a glass surface is by
chemical precipitation of metals. Iron, cobalt and nickel films may
be applied to glass surfaces by chemical precipitation methods.
Transmittance curves for different types of light- and heat-reflecting

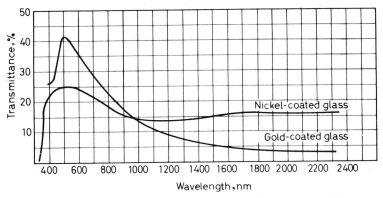

Figure 5.8. Transmittance curves for two light- and heat-reflecting glasses; nickel-coated
glass and gold-coated glass

glasses are shown in Figure 5.8. As a rule these glasses are not used
for single glazing. They are placed in sealed double-glazed units or
in laminated glass units. The metallic film is then placed on one of

47

the interior surfaces of the sealed unit. The film is therefore protected from mechanical and chemical attack. Some technical data of commercial, light- and heat-reflecting glasses are shown in *Table 9.7*.

5.7 LIGHT-SENSITIVE GLASSES

Glasses have been developed which darken when exposed to light and clear when the light source is removed (*Figure 5.9*). These light-sensitive glasses, or photochromic glasses, give variable-transmittance

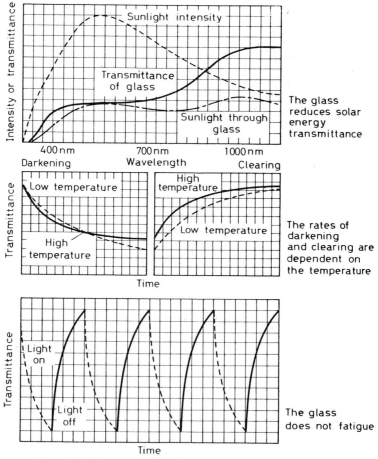

Figure 5.9. Light-sensitive glass; by courtesy of Corning Glass Works

windows. Photochromic glasses have been developed by Corning Glass Works. These glasses contain submicroscopic crystals which darken under exposure to light. The darkened colour of the glass is usually grey, sometimes tending towards brown or purple. These reactions are reversible and therefore the colour disappears when the light is removed and returns when the glass is exposed to light. The reaction to light may be almost instantaneous. The time required to regain original transmittance, however, may vary from minutes to hours. The optical transmittance of a clear photochromic glass can be equal to that of ordinary window glass. After darkening the transmittance may be as low as 1 %.

Photochromic glasses contain silver halides which give rise to the reversible reactions within the glass. They may be used as windows and sunglasses and as optical memory and self-erasing display devices.

5.8 THE DAYLIGHT FACTOR

The primary purpose of a window is to transmit natural light into a building. The two sources of this natural light are direct sunlight and sky-light. The sky can be regarded as a more effective light source than the sun and, therefore, in natural lighting problems it is customary, in most places, to consider the sky as the only source of light. It is necessary, however, to take into account the reflected sky-light as well as the direct sky-light. The amount of daylight arriving at a certain point in a building may, in fact, reach this point as direct sky-light, as externally-reflected sky-light or as internally-reflected sky-light.

The sky-light factor is equal to the ratio of the daylight falling on a surface in the building to that on a horizontal surface of the same size outside the building.

It is often of importance in the design of a building to be able to pre-estimate the daylight factors in different rooms. It is then necessary to have solar heat graphs for the site where the house is to be placed. These graphs give the heat gain due to radiation from the sun and from the sky, including reflection from the ground, through a window of a given size.

The outdoor illumination on a dull, overcast day can usually be assumed to be approximately 5000 lm/m² (465 lm/ft²). If an intensity of 600 lm/m² (56 lm/ft²) (600 lx) is required in a room, the daylight factor should be $600/5000 = 0.12$ or 12%.

Lighting intensities required for various types of work and room are given in *Table 5.2.*

The figures in *Table 5.2* are very approximate. A good indoor general lighting should probably be between 300 lx and 800 lx (28–74 lm/ft^2). In some places, however, figures as high as 1000 lx (93 lm/ft^2) are specified. These figures are steadily increasing.

Table 5.2. *Illumination values for various types of activity*

Type of room	Approximate illumination value	
	(lm/ft^2)	(lm/m^2)
Schools:		
Ordinary classroom	30 (average)	325
Laboratories	28–47	300–500
Factories:		
Ordinary work	14–28	150–300
Fine work	28–47	300–500
Very fine work	65–84	700–900
Offices		
Clerical work	28–47	300–500
Drawing office	37–47	400–500

When increasing the dimensions of a window to let in more daylight it is important to recognize that an increase in height gives a better result than an increase in width.

In comparing schools and offices it may also be borne in mind that adults need lighting intensities approximately twice as high as do children.

5.9 SOLAR SHADING DEVICES

A window glazed with two panes of plate glass will transmit a maximum heat of 600 kcal/h.m^2 (220 Btu/h.ft^2) of the sun and sky radiations. The shading effects of awnings, Venetian blinds and different kinds of sun-protecting glasses have been studied by Brown and Isfält[2]. Their results are shown in *Table 6.4.* The shading factor F_1 gives the heat transmitted by direct transmittance. F_2 is the shading factor for the total heat transmittance (direct transmittance and re-radiated heat). All figures are multiplied as to give an F_1 factor of 100 for a window glazed with two panes of sheet glass.

The total transmittance of heat through a window unit may be calculated in the following way.

$$Q = F_1 I + k(t_1 - t_2)$$

where Q is the total heat transmittance in kilocalories per square metre per hour, F_1 is the shading factor given in *Table 6.4*, I is the total heat transmittance through a double-glazed window, in kilocalories per square metre per hour, k is the heat transmittance coefficient (see *Table 6.4*) in kilocalories per square metre per hour per degree Centigrade, t_1 is the outdoor temperature and t_2 is the room temperature, in degrees Centigrade.

5.9.1 Example

Calculate the total solar heat transmittance through a double-glazed window in which there are Venetian blinds between the two panes of glass. The room temperature is 20°C (68°F) and the outdoor temperature +5°C (41°F). The total heat transmittance through a double-glazed window can be assumed to be 550 kcal/m²h (205 Btu/ft²h).

$$Q = 0.35 \times 550 + 2.1 (5 - 20)$$
$$= 193 + 10.5 - 42$$
$$= 161.5 \, \text{kcal/m}^2\text{h} \, (60 \, \text{Btu/ft}^2\text{h})$$

If the Venetian blind is placed on the room side of the window, the heat transmittance will be

$$Q = 0.56 \times 550 + 2.8 (5 - 20)$$
$$= 308 + 14 - 56$$
$$= 266 \, \text{kcal/m}^2\text{h} \, (99 \, \text{Btu/ft}^2\text{h})$$

The solar heat transmittance is thus approximately 65% higher if the Venetian blinds are put on the room side of the window than if they are placed between the two panes of glass. This indicates that solar shading devices should be placed as far out towards the external wall as possible and preferably outside the glass window.

REFERENCES AND BIBLIOGRAPHY

[1] *Glazing Manual*, Flat Glass Jobbers Association, Kansas, 1965
[2] BROWN, G. and ISFÄLT, E. 'Solskydd', *VVS*, 6 (1965)
[3] Eitel, W. and Pirani, M. *Glastechnische Tabellen*, Springer, Berlin, 1932

[4] TOLEY, F. V. *Handbook of Glass Manufacture*, Vols. 1 and 2, Ogden, New York, 1960–61
[5] PHILIPS, C. J., *Glass, its Industrial Applications*, Reinhold, New York, 1960
[6] SHAND, E. B. *Glass Engineering Handbook*, Maple Press, York, Pa., 1958
[7] MCGRATH, R. and FROST, A. C. *Glass in Architecture and Decoration*, Architectural Press, London, 1961
[8] VÖLCKERS, O. *Tafelglasdaten*, Verlag Karl Hofmann, Schorndorf bei Stuttgart, 1954
[9] SEIZ, R. *Glaser-Fachbuch*, Verlag Karl Hofmann, Schorndorf bei Stuttgart, 1963
[10] SPICKERMANN, H. *Erweitertes Gussglas Tabellarium*, Verlag Karl Hofmann, Schorndorf bei Stuttgart, 1958
[11] HOPKINSON, R. G. *Architectural Physics Lighting*, Dept. of Scientific and Industrial Research, H.M.S.O., London, 1963
[12] WIGEN, R. 'Vinduer', *Handb. Norg. Byggforskningsinst.*, 15 (1963)
[13] PLEIJEL, G. 'Solinstrålning genom Fönster', *Rapp. St. Bygnaddsforsk*, 94 (1963)
[14] JEBSEN-MARWEDEL, H. *Tafelglas*, Verlag Girardet, Essen, 1950
[15] MOREY, G. W. *The Properties of Glass*, Reinhold, New York, 1945
[16] PIGANIOL, P. *Les Industries Verrieres*, Dunod, Paris, 1966
[17] MARKUS, T. A. *Daylight with Insulation*, Pilkington Bros., Ltd., St. Helens, 1960
[18] PETHERBRIDGE, P. 'Transmission Characteristics of Window Glasses and Sun Controls', *Conference on Sunlighting in Buildings*, Newcastle upon Tyne, 1965
[19] PERSSON, R. 'Wärmeabsorbierende und Wärmereflektierende Gläser', *VDI Z.* Düsseldorf, 1 (1968)

6

THERMAL PROPERTIES

6.1 SPECIFIC HEAT

THE specific heat of a substance is defined as the quantity of heat required to raise the temperature of unit mass of the substance through one degree. The c.g.s. unit of heat is the calorie and is the quantity

Table 6.1. Thermal expansion (β), specific heat (C) and thermal conductivity (k) for some different materials

Material	$\beta \times 10^6$ per °C	per °F	C (kcal/ kg degC)*	k (kcal/ m h degC)	k (Btu in/ ft² h degF)
Acrylic plastic	90	50	0·35	0·15	1·2
Aluminium	24	13	0·22	180	1450
Brass	20	11	0·09	100	810
Cast iron	11	6	0·12	25–40	200–325
Clay brick	9–10	5–6	0·18	0·4–0·6	3·2–4·9
Concrete, dry	9–12	5–7	0·22	1·5	12
Copper	17	9	0·10	338	2730
Glass	8–8·5	4·4–4·7	0·20	0·65–0·75	5·5–7·3†
Gypsum	25	14	0·26	1·1	8·9
Lead	29	16	0·03	30	240·
Marble	12	7	0·21	2–3	16–24
Neoprene	~100	56	0·40	0·18	1·5
Oak wood					
parallel to grain	5	3	0·57	0·30	2·4
perpendicular to grain	54	30		0·14	1·1
Pine wood					
parallel to grain	5	3		0·30	2·4
perpendicular to grain	34	19		0·12	1·0
Polyvinyl chloride	70	39	0·40	0·15	1·2
Porcelain	5–8	3–4	0·20	0·90	7·3
Sandstone	7–12	4–7	0·17	1·1–2·0	9–16
Steel	12	7	0·11	40–50	325–400
Teak wood					
parallel to grain	4–6	2–3	0·40	0·32	2·6
perpendicular to grain	40–50	22–28	0·40	0·15	1·2
Zinc	29	16	0·09	97	785

 * The two units for specific heat (kcal/kg degC) and (Btu/lb degF) are identical in value since they both represent ratios of heat capacity in comparison with water.
 † For thermal calculations $k = 0.65$ and 7·3 are the values recommended.

of heat required to raise 1 g of water through 1°C. Usually a larger heat unit is used, e.g. 1 kcal = 1000 cal. The specific heats of some different building materials are shown in *Table 6.1*. The British thermal unit (Btu) is the quantity of heat required to raise the temperature of one pound of water through 1°F.

The quantity of heat necessary to raise the temperature of a substance from $t_1°$ to $t_2°$ can be calculated using the formula

$$Q = c \times m(t_2 - t_1)$$

where Q is the quantity of heat in kilocalories, or British Thermal Units, c is the specific heat in kilocalories per kilogramme per degree Centigrade or Btu per pound per degree Fahrenheit, m is the weight of the material in kilogrammes, or pounds, t_1 is the temperature at the beginning and t_2 is the final temperature, in degrees Centigrade or Fahrenheit.

6.2 METHODS OF HEAT TRANSFER

Heat always flows from a warmer to a colder body. The transfer of heat may take place by conduction, convection and radiation.

In conduction, heat passes from one part of a body to another part of the same body or to another body in direct contact with it. The rate of flow is called the thermal conductivity, or k-value, and is expressed in terms of the number of kilocalories (Btu) which will pass through one square metre (ft^2) of a material one metre (in) thick in one hour for one degree Centigrade (F) temperature difference. The heat transferred by conduction may be determined by means of the following formula

$$Q = \frac{Ak(t_2 - t_1)}{l}$$

where Q is the heat transferred by conduction in kilocalories per hour (Btu/h), A is the area perpendicular to the heat flow in square metres (square feet), k is the conductivity in kilocalories per hour per square metre per degree Centigrade, per metre thickness (Btu/ft$^2 \cdot$ h degF per in thickness), l is the thickness of the material in metres (inches) t_1 is the temperature of the hot surface in degrees Centigrade (degF) and t_2 is the temperature of the cold surface in degrees Centigrade (degF).

The conductivities of some building materials are given in *Table 6.1*.

Convection involves physical movement of the medium (usually air)

54

by which the heat is transferred. If mechanical means, such as a fan, are used for the transference the process is known as forced convection.

Thermal radiation is similar to light radiation, the essential difference being the wavelength. By radiation heat passes through space without the presence of matter.

6.3 COEFFICIENT OF THERMAL TRANSMITTANCE

6.3.1 Introduction

The overall heat transfer coefficient (thermal transmittance) of a wall is usually designated by the letter U. It is the amount of heat, in kilocalories, passing through one square metre of the wall in one hour, per degree Centigrade temperature difference between the air on the inside and outside of the wall (Btu/ft²h degF).

In calculating the U-value of a wall the following formula is used

$$U = \frac{1}{\dfrac{1}{f_i} + \sum \dfrac{l}{k} + \dfrac{1}{f_o}}$$

where l is the thickness of the material, in metres (inches), k is the conductivity in kilocalories per hour, per square metre per degree Centigrade, per metre thickness (Btu in/ft²hdegF in) and f_i and f_o are the film or surface conductance of the inside (f_i) and outside (f_o) surfaces respectively. They represent the heats expressed in kilocalories (Btu) transmitted by radiation, conduction and convection from a surface to the air surrounding it, or vice versa per square metre of surface per hour per degree Centigrade (per in²hdegF) difference between the surface and the air surrounding it. For an inside wall the surface conductance (f_i) usually equals 7 kcal/m²hdegC (1·46 Btu/ft²hdegF). For an outside wall (f_o) a value of 18 kcal/m²hdegC (6·0 Btu/ft²hdegF) is usually used. The f_o-value varies with the wind velocity. The thermal resistance of an outside wall (wind velocity 2–3 m/s) is thus

$$r_1 + r_0 = \frac{1}{7} + \frac{1}{18} \approx 0·20 \text{ m}^2\text{hdegC/kcal}$$

In British units the values $r_1 = 0·7$ and $r_0 = 0·3$ are sometimes used; $r_1 + r_0 = 0·7 + 0·3 = 1 \text{ ft}^2\text{hdegF/Btu}$. Common values in American calculations are $f_i = 1·46$ and $f_o = 6·0$ Btu/ft²h degF($f_o = 6·0$ refers to a wind velocity of 15 mph).

55

The average thermal transmittance (U-value) for glass is 5 kcal/ m²hdegC (1 Btu/ft²hdegF). It varies very little with the thickness of the glass as can be seen from Subsection 6.3.2.

6.3.2 Calculation of U-values

Using the equation in Section 6.3 the U-value of a 3 mm ($\frac{1}{8}$ in) glass pane can be calculated in the following way:

$$U = \frac{1}{\dfrac{1}{7} + \dfrac{0 \cdot 003}{0 \cdot 65} + \dfrac{1}{18}}$$

$$U = \frac{1}{0 \cdot 20 + 0 \cdot 005}$$

therefore $\qquad U = 4 \cdot 9 \text{ kcal/m}^2\text{h degC}$

or in British units

$$U = \frac{1}{0 \cdot 7 + \dfrac{\frac{1}{8}}{7 \cdot 3} + 0 \cdot 3}$$

$$U = \frac{1}{1 + 0 \cdot 017}$$

$$= 1 \text{ Btu/ft}^2\text{h degF}$$

The U-value of an insulating glass depends on the number of glass panes and the thickness of the air spaces. A double glass (3 + 3 mm) ($\frac{1}{8}$ + $\frac{1}{8}$ in) unit with a 12 mm ($\frac{1}{2}$ in) air space may be taken as an example. The thermal resistance of the sealed air space in insulating glasses can be found from *Table 6.2*.

$$U = \frac{1}{\underset{\substack{\text{(inside} \\ \text{surface)}}}{\dfrac{1}{7}} + \underset{\text{(glass)}}{\dfrac{0 \cdot 003}{0 \cdot 65}} + \underset{\text{(glass)}}{\dfrac{0 \cdot 003}{0 \cdot 65}} + \underset{\substack{\text{(sealed} \\ \text{air)}}}{\dfrac{1}{\left(0 \cdot 15 + \dfrac{2}{10} \times 0 \cdot 05\right)}} + \underset{\substack{\text{(outside} \\ \text{surface)}}}{\dfrac{1}{18}}}$$

$$U = \frac{1}{0 \cdot 20 + 2 \times 0 \cdot 005 + 0 \cdot 16}$$

$$U = \frac{1}{0 \cdot 37}$$

$U = 2{\cdot}7\,\text{kcal/m}^2\text{h degC}.$

In British units:

$$U = \frac{1}{0{\cdot}7 + 0{\cdot}017 + 0{\cdot}017 + 0{\cdot}78 + 0{\cdot}3}$$

$U = 0{\cdot}55\,\text{Btu/ft}^2\text{h degF}.$

Table 6.2. Thermal resistance of sealed air space in insulating glasses

Thickness of air space		Thermal resistance	
(mm)	(in)	m²h degC/kcal	(ft²h degF/Btu)
5	$\frac{13}{64}$	0·12	0·59
10	$\frac{25}{64}$	0·15	0·73
20	$\frac{25}{32}$	0·20	0·98
50	$1\frac{31}{32}$	0·20	0·98

6.3.3 U-values of windows

The U-values of double and triple insulating glasses are given in *Table 6.3* and shown graphically in *Figure 6.1*. The U-values decrease with increased air space up to approximately 15 mm ($\frac{5}{8}$ in). Air spaces wider than 15 mm ($\frac{5}{8}$ in) do not decrease the U-value to any appreciable extent.

Table 6.3. U-values of insulating glasses; 3 mm glass ($\frac{1}{8}$ in)

Air space		Double glass		Triple glass	
(mm)	(in)	(Btu/ft²h degF)	(kcal/m²h degC)	(Btu/ft²h degF)	(kcal/m²h degC)
4	$\frac{5}{32}$	0·68	3·35	0·49	2·40
6	$\frac{15}{64}$	0·62	3·05	0·43	2·10
8	$\frac{5}{16}$	0·59	2·90	0·40	1·95
10	$\frac{25}{64}$	0·57	2·80	0·38	1·85
12	$\frac{1}{2}$	0·55	2·70	0·37	1·80
14	$\frac{35}{64}$	0·54	2·65	0·35	1·70
20	$\frac{25}{32}$	0·53	2·60	0·34	1·65

When considering the U-value of a window it is important to know the heat-loss coefficients for the sash as well as for the glass. It is seen from *Table 6.1* that glass has a heat conductivity of 0·65

E

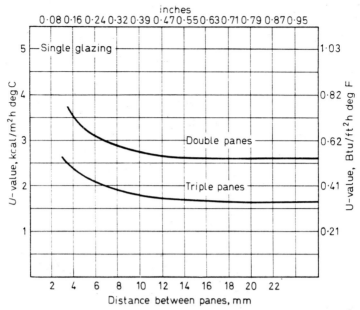

Figure 6.1. U-values of double and triple glazings with varying distances between the glass panes

kcal/mh degC (7·3 Btu in/ft²h degF). The conductivity of wood is 0·15 (1·2) (at right angles to the direction of the fibres), that of steel is 40–50 (325–400) and that of aluminium is 180 kcal/mh degC 1450 Btu in/ft²h degF). By widening a wooden sash and thereby decreasing the glass area, the U-value of the window unit will decrease. With a metal sash the U-value of the window unit will increase under similar circumstances. The combined approximate U-value of a window unit, including glass and sash, can be calculated using the following formula.

$$U_{total} = a \times U_{glass} + (1 - a) U_{sash}$$

where a is the proportion of the glass area to the total wall opening.

Figure 6.2 shows how the U-values of double glazing are influenced by the proportion of wooden and metal sashes in the window units. In cold climates it is not advisable to let a metal sash pass directly from the outside wall to the inside.

Brown and Isfält[1] have studied the heat transmittance through windows using different types of glass and different sun-protecting

58

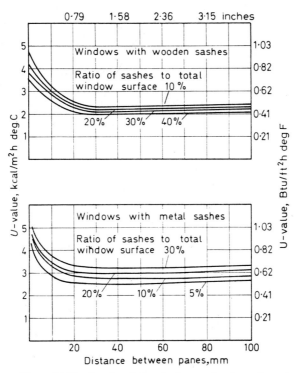

Figure 6.2. U-values for windows having wood sashes and metal sashes respectively; from W. Schüle[4] *by courtesy of* Gesund-heits-Ingenieur

devices. Their studies included heat-absorbing and heat-reflecting glasses. Their values are shown in *Table 6.4*.

Höglund[2] has compared the heat consumption in small residéntial houses in Sweden using double and triple glazing, respectively. His measurements were carried out during two different periods in 1961, period 1 from the 26th January to the 7th of March, and Period 2 from the 10th of March to the 12th of April. The results obtained by Höglund are shown in *Table 6.5*. It is interesting to notice that there is a positive heat balance when using triple glazing during the second period. The heat gain in the houses from sun and sky radiation was thus higher than the heat loss. Lindskoug[3] has shown that triple glazing may be more economical than double glazing for electrically-heated houses as far south as Rome.

5.7 LIGHT-SENSITIVE GLASSES

Table 6.4. Sun shading coefficients and heat transmittance values for different types of window. F_1 gives the total transmittance of sun energy passing the window. F_2 gives the direct transmittance. All values are related to an F_1 value of 100 for a double-glazed (two single panes) window

Type of window	F_1	F_2	U-value	
	(%)	(%)	(kcal/ m²h degC)	(Btu/ ft²h degF)
Single glazing: Ordinary glass	112	109	5·0	1·02
Double glazing:				
Ordinary glass	100	93	2·5	0·51
Heat-absorbing glass				
in outer pane	76	65	2·5	0·51
Insulating glass	31	22	2·0	0·41
outer glass heat-				
reflecting, gold coating				
Triple glazing:				
Ordinary glass	91	80	1·7	0·35
Heat absorbing glass	68	56	1·7	0·35
in outer pane				
Insulating glass, outer glass	28	20	1·5	0·31
heat reflecting, gold coating				
Double glazing: ordinary glass:				
Venetian blinds on the				
outside	9	4	2·3	0·47
Venetian blinds between	35	8	2·1	0·43
glass panes				
Venetian blinds on room side	56	8	2·8	0·57
Dark-coloured roller blind	67	0	2·2	0·45
on room side				
Triple glazing, ordinary glass:				
Venetian blinds on the outside	8	3	1·6	0·33
Venetian blinds between two	27	6	1·5	0·31
outer glass panes				
Venetian blinds between two	43	7	1·5	0·31
inner glass panes				
Venetian blinds on room side	55	7	1·8	0·37

From Brown and Isfält[1] by courtesy of Värme, Vatten Sanitet (VVS) Journal.
The U-value refers to the actual glass area only and does not include the sash.

Table 6.5. *Heat balances for double- and triple-glazed windows in residential houses in Sweden*

		Period 1		Period 2	
		(Mcal)	(Btu $\times 10^6$)	(Mcal)	(Btu $\times 10^6$)
Double glazing:	Heat loss	−574	−2·28	−450	−1·79
	Heat gain	+189	+0·75	+376	+1·49
	Total	−385	−1·53	−74	−0·30
Triple glazing:	Heat loss	−365	−1·45	−286	−1·13
	Heat gain	+170	+0·68	+337	+1·34
	Total	−195	−0·77	+51	+0·21
Saving by triple glazing		190	0·76	125	0·51

From Höglund[2] by courtesy of *Statens råd för Byggnadsforskning.*

6.4 AIR LEAKAGE

Heat transmittance calculations are usually based on the assumption that there are no infiltration losses through cracks in any part of the building.

If a window is poorly fitted, cracks may be present in certain parts along its perimeter. The volume of air that may pass through a crack depends on the size of the crack and on the wind velocity outside the building. The infiltration losses can be calculated by assuming that all air that enters has a temperature equal to the outside temperature and therefore must be heated to room temperature. The following equation may be used[4]

$$V_h = l \times a \times \Delta p^{\frac{2}{3}}$$

where V_h is the volume of air entering through the crack per hour (m³/h), l is the length of the crack in metres, a is the volume of air leakage per metre of crack at a pressure difference of 1 kgf/m² in cubic metres per hour and Δp is the difference in air pressure between the outside and inside, in kilogrammes-force per square metre, or millimetres water gauge.

The volume of air leaking through a crack varies considerably. Air leakage when using wooden frames, may often be 0·5–1·5 m³/h (18–53 ft³/h) for a crack length of 1 m (40 in) and a difference in air

61

pressure of 1 kgf/m² (0·0014 lb/in²). For metal frames the leakage may be somewhat higher, 0·5–2·5 m³/h (18–88 ft³/h). The influence of air leakage on the U-value is shown in *Figure 6.3*. It is evident from *Figure 6.3* that air leakage through cracks may substantially increase with ageing and it is therefore important to keep a close control on cracks.

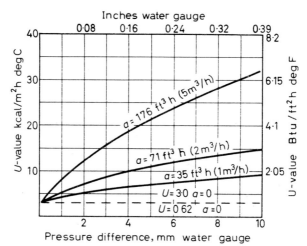

Figure 6.3. The influence on the U-value of a window of the tightness of the seal; the U-value is 3·0 kcal/m²h degC (0·62 Btu/ft²h degF) when all seals are tight, the volumes of air passing through the non-tight seals are 1, 2 and 5 m³/h (35, 71 and 176 ft³/h) respectively, the length of the seal is 4 m/m² (1·2ft/ft²); of glass surface from W. Schüle[4] by courtesy of Gesundheits-Ingenieur

It is important to see that materials used in the frame do not deteriorate. Care should be taken to give the correct treatment to wood and metal before glazing. There are many different mastics, putties and sealants used for glazing. In most cases it is advisable to use a material that does not harden with time, that has permanent elastic properties and that has good adhesive properties.

6.5 DRAUGHT FROM A COLD GLASS SURFACE

Windows, in general, have higher U-values than the surrounding wall. The coldest part of the internal wall is therefore often the glass surface. This will cause a downward draught from the window. The velocity of the air close to the glass surface may be quite high as is

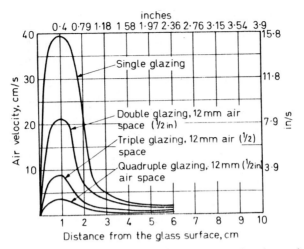

Figure 6.4. Curves showing the relationship between the velocity of the cold draught and distance from the glass surface; outdoor tempera-ture = −9°C; from H. Reiher and A. Smesny[5] by courtesy of Glastechnische Berichte

seen in *Figure 6.4*. At a distance as little as 3 cm ($1\frac{3}{16}$ in) from the glass surface, the velocity is much smaller[5]. The degree of discomfort caused by a draught depends on the temperature as well as the velocity of the cold air. A high velocity draught must be warmer than a low velocity one if discomfort is to be avoided. At a velocity as low as approximately 10 cm/s (4 in/s) a draught will be felt chilly if its temperature is lower than 18°C (64°F). The actual surface temperatures of the human body and the surrounding objects are of great importance. There should be a proper balance between internal heat production and the heat given off by the human body ($\sim$ 100 kcal/h) (400 Btu/h).

6.6 THE SURFACE TEMPERATURE OF WINDOWS

The temperature of the internal surface of single-, double- and triple-glazed windows can be found from the diagram in *Figure 6.5*. These values are only approximate since there are many factors that have an influence on the temperature of the glass surface (e.g. the U-value, the direction and velocity of the wind, etc.). Under certain conditions the water vapour in the atmosphere may condense on the internal

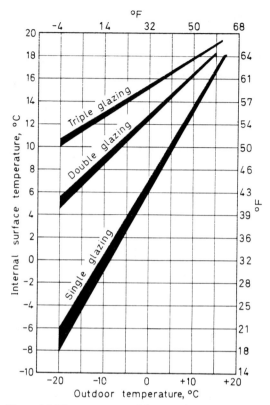

Figure 6.5. The temperature of the room-facing surface at
a room temperature of 20°C and various outdoor temperatures

surface of a window. The temperature at which condensation occurs, the dew point, varies with the humidity of the atmosphere. The dew point will increase with increased humidity.

The diagram in *Figure 6.6* shows when condensation occurs on the inside surface of a window. The following examples illustrate how to use the diagram.

(1) At what outside temperature will condensation take place on the inside surface of a double-insulating glazing unit (U-value = $2\cdot70$ kcal/m^{2}h degC) ($0\cdot55$ Btu/ft^{2}h degF), when the room temperature is 20°C (68°F) and the relative humidity 40%?

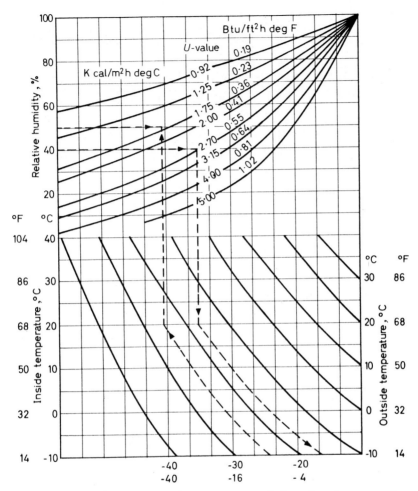

Figure 6.6. Diagram for determining the temperature at which condensation occurs; from DIN 4701 : 1959

Start at 40% relative humidity and follow this line to the curve for U = 2·70 (0·55). Draw a vertical line to 20°C (68°F) of room temperature. Draw a curve parallel to the adjacent curves for outside temperature and read off −17°C (+1°F) which is the required answer.

(2) What U-value is required for a window in order to prevent

condensation at an outside temperature of $-25°C$ ($-13°F$)? The room temperature is $20°C$ ($68°F$) and the relative humidity 50%.

Start at $-25°C$ ($-13°F$) outside temperature. Draw a curve to $20°C$ ($68°F$) room temperature. Draw a straight line vertically to the ordinate for 50% relative humidity. The U-value can be estimated to be 1·7 (0·35).

6.7 THERMAL EXPANSION

The linear coefficient of expansion for flat glass, per unit change in temperature (°C) is 80–85 (44–47) $\times$ 10^{-7}. This is the same whether the unit of length is the centimetre, foot or inch. A light of glass 1 m (= 1000 mm) long will thus expand 0·80–0·85 mm for a 100°C increase in temperature, or roughly 1 mm per metre. (A 1000 in glass edge will increase 0·44–0·47 in for 100°F increase.) This is a comparatively small thermal expansion compared with most other building materials as seen from *Table 6.1*. The expansion of steel is about 50% higher than that of glass and aluminium expands three times as much as glass. This must be taken into consideration when glazing in metal sashes.

Due to its brittle properties glass cannot withstand thermal shocks as well as most other materials. Thin glass is more resistant than thick glass in this respect. *Table 6.6* gives some approximate figures

Table 6.6

Thickness of glass		Maximum temperature difference it can withstand	
(mm)	(in)	(degC)	(degF)
2	$\frac{3}{32}$	100	180
3	$\frac{1}{8}$	90	160
4	$\frac{5}{32}$	80	140
6	$\frac{1}{4}$	60	110

for the maximum thermal shock of sheet glass of different thicknesses[6].

In general, it can be said that glass can withstand twice as high a heating shock as a cooling shock. The reason for this is that compressive stresses develop during heating while tensile stresses will be set up on the surface during cooling.

The resistance to thermal shock is higher for tempered glass. In general this glass can withstand a shock of 300°C (540°F).

REFERENCES AND BIBLIOGRAPHY

[1] BROWN, G. and ISFÄLT, E. 'Solskydd', *VVS*, 6 (1965)

[2] HÖGLUND, I. 'Värmförluster i Småhus', *Rapp. St. Byggnadsforsk*, **43** (1963)

[3] LINDSKOUG, N. E. 'Views on Electrical Space Heating', *HSB*, Dept. of Building Research, Stockholm, 1964

[4] SCHÜLE, W. 'Untersuchungen über die Luft- und Wärmedurchlässigkeit von Festern', *Gesundheits-Ingenieur*, 6 (83) (1962) 153

[5] REIHER, H. and SMESNY, A. 'Beitrage zur Frage der "Kaltluft-Kaskade" an Festern,' *Glastech Ber.* 10 (37) (1964) 476

[6] JEBSEN-MARWEDEL, H. 'Tafelglas,' Verlag Girardet, Essen, 1950

[7] EITEL, W. and PIRANI, M. *Glastechnische Tabellen*, Springer, Berlin, 1932

[8] TOLEY, F. V. *Handbook of Glass Manufacture*, Vols. 1 and 2, Ogden, New York, 1960–61

[9] PHILIPS, C. J. *Glass, its Industrial Applications*, Reinhold, New York, 1960

[10] SHAND, E. B. *Glass Engineering Handbook*, Maple Press, York, Pa., 1958

[11] MCGRATH, R. and FROST, A. C. *Glass in Architecture and Decoration*, Architectural Press, London, 1961

[12] VÖLCKERS, O. *Tafelglasdaten,* Verlag Karl Hofmann, Schorndorf bei Stuttgart, 1954

[13] SEIZ, R. *Glaser-Fachbuch*, Verlag Karl Hofmann, Schorndorf bei Stuttgart, 1963

[14] Spickermann, H. *Erweitertes Gussglas Tabellarium*, Verlag Karl Hofmann, Schorndorf bei Stuttgart, 1958

[15] Hopkinson, R. G. *Architectural Physics Lighting*, Dept. of Scientific and Industrial Research, H.M.S.O., London, 1963

[16] GJELSVIK, T. 'Tests with Factory-sealed, Double-glazed Window units,' *Rapp. Norg. ByggsforskInst.* **33** (1962)

[17] Clore, P. D. *Sound Control and Thermal Insulation*, Reinhold, New York, 1966

[18] MOREY, G. W. *The Properties of Glass*, Reinhold, New York, 1945

[19] PIGANIOL, P. *Les Industries Verrieres*, Dunod, Paris, 1966

[20] SCHAUPP, W. *Die Aussenwand*, Verlag Georg Callway, Munich, 1962

[21] MARKUS, T. A. *Daylight with Insulation*, Pilkington Bros. Ltd., St. Helens, 1960

[22] PERSSON, R. *Wärmeabsorbierende und Wärmereflektierende Gläser, VDI Z.,* Düsseldorf, 1 (1968)

7

ACOUSTICAL PROPERTIES

7.1 DEFINITIONS

SOUND is produced by vibrating bodies which give rise to wave motions in the air. These sound waves expand outwards in all directions from the vibrating body which can be assumed to vibrate about its equilibrium position. The extreme displacement on either side of the equilibrium position is called the amplitude of the vibration. The number of complete vibrations made by the body in one second is called the frequency of the vibrating body. The frequency is measured in hertz (Hz), where 1 Hz is one cycle or complete vibration per second. The frequency of audible sounds varies from 20 Hz to 20,000 Hz. In acoustical measurements the most commonly used range is 100–4000 Hz. The sound level or intensity is expressed in the decibel unit (dB). Hearing is a primary sensation caused by a stimulation of the auditory nerve of the ear by sound waves. *Table 7.1* gives some average sound levels.

It is usual to distinguish between air-borne and impact sounds. The air-borne sound is transmitted through wall, floor or partition by means of a diaphragm action and thus carried on into the adjacent

Table 7.1. Some average sound levels

	Intensity (dB)
Limit of audibility	0–15
Radio studio without public	15–30
Auditorium	20–40
Apartments, schools and offices	30–50
Average conversation	40–50
Retail·stores	60–90
Average factory	70–90
Noisy street	90–100
Noisy factory	100–120
Threshold of hearing	115–170
Close to a jet engine	120–140

room. Impact sounds on the other hand are produced by footsteps, door-slams or other mechanical impacts and these sounds are structure-borne.

The sound insulation value of a structure, e.g. a wall or a window is usually expressed in terms of transmission losses (TL). The 'transmission loss' of a structural unit is the number of decibels by which the level of air-borne sound is decreased when passing through this structure. A window with a transmission loss of 35 dB will reduce a sound intensity from the street of 90 dB to (90–35) = 55 dB, in a room, provided there are no other sound paths to the room.

7.2 SOUND INSULATION BY SINGLE GLAZING

A single glass of ordinary thickness has a relatively low sound insulation value, as seen from the values given in *Table 7.2*. These values vary to some extent with the dimension of the glass pane.

Table 7.2

Glass thickness		Insulation value
(mm)	(in)	(dB)
2	$\frac{3}{32}$	26–27
4	$\frac{5}{32}$	27–31
6	$\frac{1}{4}$	28–32
8	$\frac{5}{16}$	29–33
12	$\frac{1}{2}$	34–37
25	1	38–42

7.3 SOUND INSULATION BY DOUBLE AND TRIPLE GLAZING

The improvement in sound insulation by double and triple glazing compared to that by single glazing is mainly dependent on the air space between the panes of glass. The sound insulation will increase with increased air space up to approximately 100 mm (4 in). Some results that have been obtained by Brandt[1] on Swedish windows are shown in *Figure 7.1*.

It is seen from these curves that a double-glazed window with a wide air space may give a better sound insulation value than a triple-glazed window with a narrow air space. In some cases it is

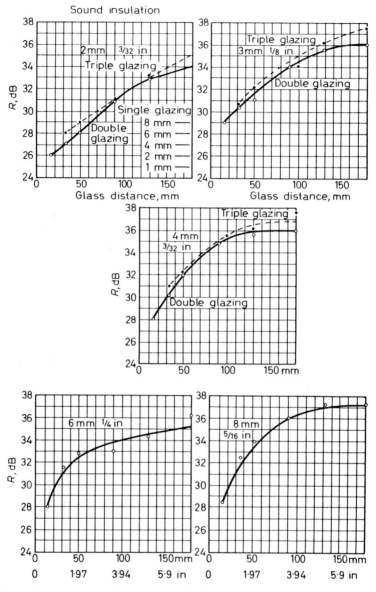

Figure 7.1. *Average sound insulation for double and triple glazings of different glass substances and distances between the glass panes; for triple glazing the distance measured is that between the two outer glass panes; from* O. Brandt[1] *by courtesy of* Statens Nämnd för Byggnadsforskning

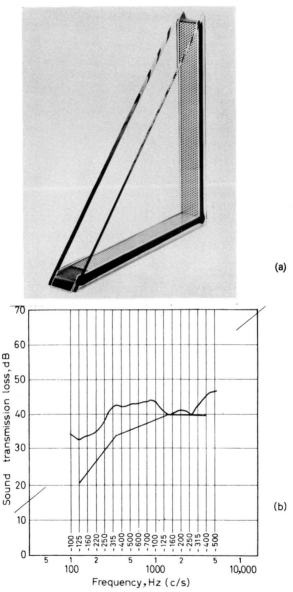

Figure 7.2. (a) Sound-insulating glass unit; (b) graph of sound transmission loss against frequency; nine frequency average, 40 dB; sound transmission class (s.t.c.) 40; by courtesy of Polarpane Corporation

71

possible to get a lower insulation value in a double-glazed window when using thick glass than one does when using thinner glass [compare the curves for 4 and 6 mm ($\frac{5}{32}$ and $\frac{1}{4}$ in) glasses respectively]. The reason for this is that the glass, which is itself vibrating, may

Table 7.3. Sound transmission loss tests carried out by the Riverbank Acoustical Laboratories, Geneva, Illinois, U.S.A. for Polarpane Corporation, Pennsauken, New Jersey

Frequency	Transmission loss	Transmission loss
(Hz)	(dB)	(dB)
125	32	25
175	34	31
250	38	36
350	42	40
500	43	42
700	43	44
1 000	44	45
1 400	(40)	(44)
2 000	41	39
2 800	(41)	(36)
4 000	46	42
Sound transmission class	40	38

Table 7.4. Sound insulation values

Description	Insulating value*
	(dB)
Single sheet of 0·6 mm (0·024 in) aluminium	16†
Single sheet of 0·7 mm (0·028 in) galvanized iron	32†
Single sheet of 3 mm ($\frac{1}{8}$ in) 3-ply plywood	26†
Single sheet of 12 mm ($\frac{1}{2}$ in) wood fibreboard	22†
Single sheet of 3 mm ($\frac{1}{8}$ in) sheet glass	28†
Single sheet of 6 mm ($\frac{1}{4}$ in) plate glass	32†
Partition of 9·2 × 12 × 19·6 mm (0·36 × $\frac{1}{2}$ × 0·77 in) of glass bricks	41†
Sound insulating window	
Gypsum wallboard (16 mm) ($\frac{5}{8}$ in), applied vertically and laminated over 12 mm ($\frac{1}{2}$ in) sound deadening board nailed vertically 305 mm o.c. to both sides of staggered 2 × 4 studs 405 mm (16 in) o.c. on 2 × 6 plates	49‡

* Nine frequency average (128–4096 Hz).
† From United States Department of Commerce, National Bureau of Standards, Building Materials of Structure, Report 144, 1955.
‡ From Clove[2] by courtesy of Reinhold Publishing Corporation.

create sound waves which increase the sound level (effect of co-incidence). A somewhat higher sound reduction value may be obtained by using panes of different thicknesses in a double-glazed window. This will preclude the possibility of resonant vibration. If a triple-glazed window and a double-glazed window have similar total thicknesses their sound insulation values will be almost identical. There is thus no advantage in having a third pane of glass unless the total thickness of the window is increased.

Some advantage may be obtained by having a sound-absorbing material at the edges between the glass panes. In properly designed sound-insulating windows it is possible to get insulating values higher than 40 dB.

It is very important to thoroughly seal the windows from the frames. This may be done by caulking or by means of gaskets of rubber.

Some acoustical data from a specially-designed sound control glass unit are shown in *Table 7.3*.

REFERENCES AND BIBLIOGRAPHY

[1] BRANDT, O. 'Akustisk' Planering' *St. Nämnd Byggnforsk Handb.* 1, Stockholm, 1958
[2] CLOVE, P. D. *Sound Control and Thermal Insulation,* Reinhold, New York, 1966
[3] EITEL, W. and PIRANI, M. *Glastechnische Tabellen*, Springer, Berlin, 1932
[4] PHILIPS, C. J. *Glass, its Industrial Applications*, Reinhold, New York, 1960
[5] SHAND, E. B. *Glass Engineering Handbook*, Maple Press, York, Pa., 1958
[6] McGRATH, R. and FROST, A. C. *Glass in Architecture and Decoration*, Architectural Press, London, 1961
[7] VÖLCKERS, O. *Tafelglasdaten,* Verlag Karl Hofmann, Schorndorf bei Stuttgart, 1954
[8] SCHNECK, A. G. *Fenster,* Julius Hofmann Verlag, Stuttgart, 1963
[9] JEBSEN-MARWEDEL, H. *Tafelglas,* Verlag Girardet, Essen, 1950
[10] PIGANIOL, P. *Les Industries Verrieres,* Dunod, Paris, 1966
[11] MARKUS, T. A. *Daylight with Insulation*, Pilkington Bros. Ltd., St. Helens, 1960

F

8

ELECTRICAL PROPERTIES

FOR all practical purposes glass can be regarded as a non-conductor of electric current at room temperature. However, the electrical resistivity of glass decreases with increased temperature. The change of its electrical resistance with temperature follows an exponential equation over a wide temperature range.

$$p = Be^{A/T}$$

where p is electrical resistivity and B and A are constants.

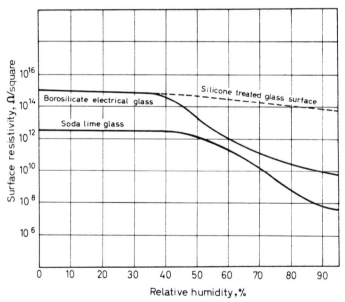

Figure 8.1. *The surface resistivity of glass as a function of relative humidity; the resistivity may be increased by using a special type of glass, e.g. borosilicate glass, or by treating the surface with silicone; by courtesy of* Corning Glass Works

The conduction is ionic in character. In this respect glass can be regarded as an electrolyte in which the sodium ion is the carrier of the current. At high temperature glass is a good conductor of electricity. If the resistivity is $10^{12}\Omega$. cm at room temperature it may be as low as 5–50 Ω. cm at 1400°C. Due to this high conductivity of hot glass it is possible to use electric melting methods in glass furnaces. Molybdenum electrodes may be immersed in the glass bath.

At high humidities the surface of a glass may be corroded. This may result in a continuous conducting layer being formed on the glass surface. Even if there is no conduction of electric current through glass at room temperature there may thus be some surface conductance if a glass has been stored in humid conditions (*Figure 8.1*). The following are some electrical data for glass at 20°C (68°F).

Dielectric constant	6–7
Resistivity	10^{11}–10^{12} Ω. cm
Dielectric power factor tg σ, 50 Hz	0·005–0·01
Dielectric strength in air at 50 Hz	300–800 kV/cm

BIBLIOGRAPHY

EITEL, W. and PIRANI, M. *Glastechnische Tabellen*, Springer, Berlin, 1932

McGRATH, R. and FROST, A. C. *Glass in Architecture and Decoration*, Architectural Press, London, 1961

MOREY, G. W. *The Properties of Glass*, Reinhold, New York, 1945

PHILIPS, C. J. *Glass, its Industrial Applications*, Reinhold, New York, 1960

PIGANIOL, P. *Les Industries Verrieres*, Dunod, Paris, 1966

SHAND, E. B. *Glass Engineering Handbook*, Maple Press, York, Pa., 1958

TOLEY, F. V. *Handbook of Glass Manufacture*, Vols. 1 and 2, Ogden, New York, 1960–61

VÖLCKERS, O. *Tafelglasdaten*, Verlag Karl Hofmann, Schorndorf bei Stuttgart, 1954

9

FLAT GLASS PRODUCTS

9.1 DEFINITIONS

FLAT glass may be manufactured by drawing, rolling, casting or blowing. The flat glass thus obtained may be further processed to make various flat glass products. The manufacturing methods will differ from one factory to another but the products made are usually quite similar and it is therefore possible to use similar definitions for them.

In many countries there are special standard specifications for flat glass products. The discussions in this chapter will be based mainly on British Standard Specifications and American Federal Specifications. A reference list of some national standards is given in Appendix II. Most national standards give definitions of the different flat glass products as well as certain quality specifications.

The flat glass products may be classified in the following way:
(1) Clear sheet glass
(2) Polished plate glass
(3) Float glass
(4) Rolled glass (cast glass)
    (a) Rough cast glass
    (b) Figured, rolled glass
    (c) Cathedral glass
    (d) Wired glass
    (e) Rolled opal glass
(5) Hand-blown sheet glass (antique glass)
(6) Heat-absorbing glass
(7) Light- and heat-reflecting glass
(8) Processed glass
    (a) Toughened glass
      (i) Tempered or heat-treated glass
      (ii) Chemically-treated glass
    (b) Laminated glass
    (c) Sealed, insulating glass units

(d) Cladding glass
(e) Sandblasted glass
(f) Acid-etched glass
(g) Metallized glass, mirrors
(h) Bent glass
(i) Edge work and bevelling

9.2 CLEAR SHEET GLASS

Almost all sheet glass is made by drawing the glass directly from the furnace into a flat sheet. The two surfaces have been fire-finished in the drawing process and are therefore plane and smooth. However,

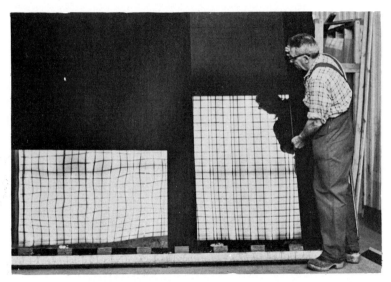

Figure 9.1. Quality control of sheet glass; by projecting light through a screen it is possible to study the flatness and optical homogeneity of the glass pane; the sample on the right is a B quality and that to the left a greenhouse quality; by courtesy of Oxelösunds Järnverk

they are never perfectly flat and parallel, there is a characteristic waviness of the surface which results in some distortion of vision and reflection.

Sheet glass is usually available in three or four different standard qualities. British Standard 952:1964 gives the following classification:

(a) *Special selected quality (S.S.Q.)*. For high-grade work where a superfine sheet glass is required, e.g. pictures, cabinet work etc.

(b) *Selected glazing quality (S.Q.)*. For glazing work requiring a selected sheet glass above the ordinary glazing quality.

(c) *Ordinary glazing quality (O.Q.)*. For general glazing purposes, e.g. for factories, housing estates, etc.

(d) *Horticultural quality*. An inferior quality, available in limited sizes for horticultural purposes.

In the American Federal Specification DD-G-451a the four qualities are referred to as AA, A, B and greenhouse qualities. The following specific requirements for these qualities are given:

AA quality. The defects permitted will be only the slightest or minute imperfections and must be almost imperceptible. Occasional seeds up to $\frac{1}{16}$ in (1·6 mm) in the central area and not exceeding $\frac{1}{8}$ in (3·2 mm) in the outer area will be permitted.

Figure 9.2. Hand cutting of sheet glass; by courtesy of Oxelösunds Järnverk

A quality. The defects permitted in the central area of this quality are a few seeds, an occasional large seed not more than $\frac{1}{4}$ in long, faint strings or lines, and very light scratches and other surface defects detected only by close scrutiny. No light shall contain all of these defects and those present may not be clustered when in the central area. In general, the central area of a light shall be practically free from defects, and the appearance of the light as a whole shall be such that there is no perceptible interference with vision as long as one is looking through the glass at an angle of not less than 45° to the surface of the glass.

B quality. The defects permitted are more prominent than permitted in A quality, for example scattered blisters not more than $\frac{1}{2}$ in long in the central area of the light, some seeds (more prominent than permitted in A quality), lines, dirt specks, and other surface defects that do not perceptibly interfere with vision normal to the glass. The outer area may contain blisters not exceeding $\frac{3}{4}$ in in

Table 9.1. *Thickness of sheet glass*

Thickness		Weight per unit surface area	Commercial thickness designation	Thickness		Weight per unit surface area	Commercial thickness designation
(mm)	(in)	(oz/ft²)		(mm)	(in)	(oz/ft²)	
0·97	0·038	8	Thin	3·78	0·149	31	
1·10	0·043	9		3·90	0·154	32	32 oz
1·22	0·048	10	10 oz	4·02	0·158	33	
1·34	0·053	11		4·14	0·163	34	
1·46	0·058	12	12 oz	4·26	0·168	35	
1·58	0·062	13		4·38	0·172	36	
1·71	0·067	14	14 oz	4·51	0·178	37	
1·83	0·072	15		4·63	0·183	38	
1·95	0·077	16	16 oz	4·76	0·188	39	$\frac{3}{16}$ in
2·07	0·082	17		4·87	0·192	40	
2·19	0·086	18	18 oz	4·99	0·196	41	
2·31	0·091	19	S.S.	5·12	0·202	42	
2·44	0·096	20		5·24	0·206	43	
2·56	0·101	21		5·36	0·211	44	
2·68	0·106	22		5·48	0·216	45	
2·80	0·110	23		5·56	0·219	46	$\frac{7}{32}$ in
2·92	0·115	24	24 oz	5·72	0·225	47	
3·04	0·120	25	D.S.	5·85	0·230	48	
3·17	0·125	26	26 oz	5·97	0·235	49	
3·29	0·130	27		6·09	0·240	50	
3·41	0·134	28		6·21	0·245	51	$\frac{1}{4}$ in
3·58	0·141	29		6·35	0·250	51	
3·65	0·144	30					

length as well as the defects permitted in the central area. With all these defects as a combination, there shall not be the resemblance of coarse grading. The appearance of the light as a whole shall be such that there is no perceptible interference with vision as long as one is looking through the glass at an angle of 90° to the surface of the glass.

Greenhouse quality. The glass shall contain no stones which may cause spontaneous breakage.

Table 9.2. Limits for sheet glass

Nominal substance or thickness		Limits of thickness	
	(mm)	(in)	(mm)
24 oz	2·9	0·108–0·120	2·75–3·05
26 oz	3·2	0·122–0·138	3·1–3·5
32 oz	3·9	0·153–0·169	3·9–4·3
$\frac{3}{16}$ in	4·7	0·183–0·207	4·65–5·25
$\frac{7}{32}$ in	5·6	0·209–0·228	5·3–5·8
$\frac{1}{4}$ in	6·2	0·246–0·266	6·25–6·75

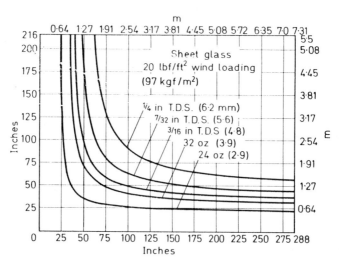

Figure 9.3. Glazing sizes for sheet glass at normal wind load; by courtesy of Pilkington Brothers Limited

Example—24 oz sheet glass; 41 × 41 in is the largest square pane which is adequate. A rectangular pane 72 × 29 in is also suitable.

Sheet glass is normally available in thicknesses ranging from 2 mm to 7 mm ($\frac{1}{16}$ in–$\frac{9}{32}$ in), however, for special purposes, other thicknesses may be manufactured. A 'single strength' (S.S.) sheet glass is 2·3 mm ($\frac{3}{32}$ in) thick and a double strength (D.S.) glass is 3·0 mm ($\frac{1}{8}$ in) thick. Instead of giving the thickness of a sheet glass its weight per unit surface area may be given (oz/ft^2). *Table 9.1* summarizes the different measurements of normal sheet glass.

A thickness tolerance of approximately 10% must be allowed for sheet glass. in BS 952:1964 the limits for sheet glass are as given in *Table 9.2*.

When determining what thickness of sheet glass to choose various factors have to be considered (e.g. glazing size, height of building, wind velocities, etc). *Figure 9.3* gives some examples of recommended glazing sizes.

9.3 POLISHED PLATE GLASS

Plate glass is usually made in a continuous sheet by a horizontal rolling process. It may also be made by casting and rolling large sheets separately. Both surfaces of the sheet are then ground and polished. The surfaces of polished plate glass are flat and parallel. The glass provides clear undistorted vision and reflection.

B.S. 952:1964 defines three different qualities:

(1) *Silvering quality (S.Q.)*. For high-grade mirrors and for all purposes where a superfine glass is required.

(2) *Selected glazing quality (S.G.)*. For better-class work and also for mirrors and bevelling.

(3) *Glazing quality for glazing (G.G)*. Standard for general glazing purposes.

The American Federal Specification, DD-G-451a defines the three different qualities, silvering quality, mirror glazing quality and glazing quality in the following way:

The variation in thickness shall be not more than $\frac{1}{32}$ in (0·8 mm) for individual lights of area under 10 ft^2 (0·9 m^2) for thicknesses up to and including $\frac{1}{4}$ in (6·4 mm).

Specific requirements for different sizes in nominal thicknesses of $\frac{1}{4}$ in or less. Type I plate glass $\frac{1}{4}$ in or less in nominal thickness shall be furnished in silvering, mirror glazing, and glazing qualities, as specified. As allowable tolerances in quality vary considerably with the size of sheet required, different requirements will apply in each of the following four divisions according to size; division I, sheets up to an including 10 ft² (0·9 m²) in area; division II, sheets having an area greater than 10 ft² (0·9 m²) but not greater than 25 ft² (2·3 m²); division III, sheets having an area greater than 25 ft²(2·3 m²) but not greater than 75 ft² (7 m²); division IV, sheets having an area greater than 75 ft² (7 m²).

Figure 9.4. Packaging end of a plate glass line; by courtesy of Pittsburgh Plate Glass Company

Division I (sizes up to and including 10 ft² (0·9 m²), nominal thickness ¼ in (6.4 mm) or less).

Silvering quality. This quality shall not contain any major defects. The central area of this quality may contain well-scattered fine seed and faint hairlines when not grouped, and occasional very light short finish visible only upon close inspection. The outer area, in addition, may contain seed and short sleeks when not grouped.

Mirror glazing quality. The central area of this quality may contain scattered seed, faint hairlines, and light short finish. The outer area, in addition, may contain light scratches and occasional faint strings not over 2 in (51 mm) long. The outer area may contain coarse seeds and an occasional bubble not larger than $\frac{3}{64}$ in (1·2 mm) in diameter.

Glazing quality. The central area of this quality may contain short finish, scattered seeds, including an occasional coarse or open seed, but no heavy seeds. Light scratches that can be removed by buffing will be permitted. The outer area, in addition, may contain bubbles not larger than $\frac{1}{16}$ in (1·6 mm) in diameter and open bubbles not larger than $\frac{3}{64}$ in (1·2 mm) in diameter, faint strings, and light ream. Stones, large bubbles, skim, pronounced ream, or heavy scratches will not be permitted in either the central or outer area.

Division II (*sizes greater than 10 ft²* (*0·9 m²*) *but not greater than 25 ft²* (*2·3 m²*), *nominal thickness* $\frac{1}{4}$ *in* (*6·4 mm*) *or less*).

Silvering quality. The central area of this quality may contain more numerous fine seeds than the small sizes, faint hairlines, and occasional light short finish. The outer area may contain occasional coarse seed and short faint scratches when not grouped.

Mirror glazing quality. The central area of this quality may contain more numerous fine seeds and light short finish than the small sizes and an occasional coarse seed. The outer area may contain coarse seeds, an occasional bubble not larger than $\frac{3}{64}$ in (1·2 mm) in diameter, an occasional small open seed, light ream, and fine strings. Heavy defects or scratches which cannot be removed by buffing will not be permitted. The polish must be good, but more pronounced short finish may be present in the outer area.

Glazing quality. The central area may contain small bubbles not exceeding $\frac{1}{16}$ in (1·6 mm) in diameter if not grouped and occasional open bubbles not exceeding $\frac{3}{64}$ in (1·2 mm) in diameter and fine strings or ream which do not give visible distortion when looking straight through the glass, but no heavy scratches. Light scratches which may be removed by buffing will be permitted. The outer area may contain bubbles $\frac{3}{32}$ in (2·4 mm) in diameter and open bubbles not over $\frac{1}{16}$ in (1·6 mm) in diameter, visible scratches which may be removed by buffing, light ream, strings, and small stones not larger than $\frac{1}{32}$ in (0·8 mm), but these defects should not be grouped or interfere with the vision. The polish over the central area should be good, but more pronounced short finish may be present in the outer area.

Division III (*sizes greater than 25 ft²* (*2·3 m²*) *but not greater than 75 ft²* (*7 m²*) *nominal thickness* $\frac{1}{4}$ *in* (*6·4 mm*) *or less*). Polished plate glass in division III shall be of glazing quality only. Lights of this size may contain more numerous and larger imperfections than are allowed in the smaller lights, but these must not be grouped or so prominent that they noticeably interfere with the vision. The central

83

area of the plate shall be free from these larger defects. The lights may contain seeds of any size but not heavy seed except in relatively small patches on the outer area of the sheet. The lights may contain occasional bubbles up to $\frac{1}{8}$ in (3·2 mm) and open bubbles up to $\frac{3}{64}$ in (1·2 mm) in the central area and occasional bubbles up to $\frac{3}{16}$ in (4·8 mm) and open bubbles up to $\frac{1}{16}$ in (1·6 mm) in the outer area. Strings, ream, and skim in very limited areas, if not causing a deformation of objects viewed through the plate, occasional scratches, and small stones under $\frac{1}{16}$ in (1·6 mm) will be permitted. Heavy ream, heavy cords, bubbles larger than $\frac{3}{16}$ in (4·8 mm) in diameter, stones larger than $\frac{1}{16}$ in (1·6 mm) in diameter, large fire cracks, areas of unpolished glass, easily visible short finish, and large open bubbles will not be permitted. The large defects shall be confined to the outer area of the sheet; the central area shall be relatively free from major defects.

Division IV (sizes greater than 75 ft² (7 m²), nominal thickness $\frac{1}{4}$ in (6·4 mm) or less). Polished plate glass in division IV shall be of glazing quality only. Sheets larger than 75 ft² (7 m²) may contain defects of almost any kind except that they must not show large areas of numerous bubbles, have any defects which wlll cause spontaneous breakage such as skim or large stones (over $\frac{1}{8}$ in (3·2 mm) in diameter), or show any areas of unpolished glass or large areas of easily visible short finish.

Specific requirement for glass $\frac{3}{8}$ in (9·5 mm) and over in nominal thickness. Polished plate glass $\frac{3}{8}$ in (9·5 mm) and over in nominal

Table 9.3. Limits for clear plate glass

Nominal thickness		Limits of thickness	Normal maximum size	
(in)	(mm)	(in)	(in)	(m)
$\frac{1}{8}-\frac{3}{16}$	3·18–4·76	0·125–0·188	90 × 50	2·3 × 1·3
$\frac{3}{16}$	3·97–5·56	0·156–0·219	100 × 72	2·5 × 1·8
$\frac{1}{4}$ bare	4·76–6·35	0·188–0·25	100 × 72.	2·5 × 1·8
$\frac{1}{4}$	5·56–7·94	0·219–0·312	175 × 98	4·5 × 2·5
$\frac{1}{4}$ exact	5·95–6·75	0·234–0·266	100 × 72	2·5 × 1·8
$\frac{5}{16}-\frac{3}{8}$	7·94–9·02	0·312–0·391	180 × 130	4·6 × 3·3
$\frac{3}{8}$	9·13–10·72	0·359–0·422	280 × 130	7·1 × 3·3
$\frac{1}{2}$	11·91–13·49	0·469–0·531	156 × 96	4·0 × 2·4
$\frac{5}{8}$	15·08–16·67	0·594–0·656	156 × 96	4·0 × 2·4
$\frac{3}{4}$	18·25–19·84	0·719–0·781	156 × 96	4·0 × 2·4
$\frac{7}{8}$	21·43–23·02	0·844–0·906	144 × 96	3·7 × 2·4
1	24·61–26·19	0·969–1·031	144 × 96	3·7 × 2·4

thickness, shall be furnished in one quality only. It may contain proportionately more defects than $\frac{1}{4}$ in (6·4 mm) glazing quality plate glass of the same sizes to compensate for the increased glass thickness over $\frac{1}{4}$ in (6·4 mm).

The range of thicknesses given in BS 952: 1964 are shown in *Table 9.3.*

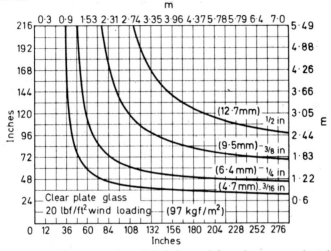

Figure 9.5. Glazing sizes for polished plate and float glass at normal wind loading; by courtesy of Pilkington Brothers Limited

Example —$\frac{1}{4}$ in substance 78 × 78 in is the largest square pane which is adequate. A rectangular pane 144 × 54 in is also suitable.

The three plate glass (and float glass) qualities are sometimes called VA, VVA and VVV respectively. VA is the highest quality and VVV is the ordinary glazing quality. Since the beginning of 1967 many manufacturers make only two qualities—a VVA (or English S.G.) silvering quality and a VVV (or English G.G.) glazing quality. Thicknesses and tolerances for both plate and float glasses are as given in *Table 9.4.*

Table 9.4

Thickness		Tolerance		Thickness		Tolerance	
(mm)	(in)	(mm)	(in)	(mm)	(in)	(mm)	(in)
3	$\frac{1}{8}$	±0·2	± 0·008	8	$\frac{5}{16}$	±0·3	± 0·012
4	$\frac{5}{32}$	±0·2	± 0·008	10	$\frac{3}{8}$	±0·3	± 0·012
5	$\frac{3}{16}$	±0·2	± 0·008	12	$\frac{1}{2}$	±0·3	± 0·012
6	$\frac{1}{4}$	±0·2	± 0·008	15	$\frac{19}{32}$	±0·3	± 0·012

9.4 FLOAT GLASS

Float glass is made by the continuous float glass process developed by Pilkington Brothers Ltd., England. The quality of float glass is similar to that of plate. BS 952:1964, therefore, gives the same quality standards for polished plate glass and float glass. The information given in *Tables 9.3* and *9.4* and *Figure 9.5* for plate glass applies equally well to float glass.

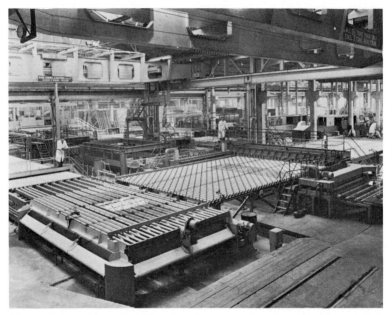

Figure 9.6. Automatic cutting of float glass; by courtesy of Pilkington Brothers Limited

9.5 ROLLED GLASS

Most rolled glass is made by a horizontal, continuous rolling process (*Figure 9.7*). For some special types of glass intermittent casting and rolling processes may be used.

9.5.1 Rough cast glass

One surface of a rough cast glass has a definite texture and the other surface is usually flat. This glass can be regarded as an unpolished plate glass. Its thickness may vary from 5 mm to 25 mm ($\frac{3}{16}$–1 in). Rough cast glass is usually supplied in one quality only.

Figure 9.7. An annealing lehr for rolled glass; the lehr is equipped with an automatic temperature regulating system; by courtesy of Societe d'Etudes et Installations Industrielles, Brussels

Thicknesses and normal maximum sizes, as given in BS 952:1964, are shown in *Table 9.5.*

Table 9.5. Limits of rough cast glass

Nominal thickness	Limits of thickness		Normal maximum size	
(in)	(mm)	(in)	(m)	(in)
$\frac{3}{16}$	4·75–5·5	0·187–0·217	3·7 × 1·2	146 × 48
$\frac{1}{4}$	6·35–7·0	0·250–0·276	3·7 × 1·2	146 × 48
$\frac{3}{8}$	9·5–10·3	0·374–0·512	3·7 × 1·2	146 × 48
$\frac{7}{16}$	9·5–11·5	0·374–0·453	4·6 × 3·3	180 × 130
$\frac{1}{2}$	11·3–13·5	0·453–0·532	7·1 × 3·3	280 × 130
$\frac{5}{8}$	14·5–16·5	0·571–0·650	4·0 × 2·4	156 × 96
$\frac{3}{4}$	18·5–20·5	0·728–0·807	4·0 × 2·4	156 × 96
$\frac{7}{8}$	21·5–23·5	0·846–0·925	4·0 × 2·4	156 × 96
1	24·5–26·5	0·965–1·043	3·7 × 2·4	144 × 96

9.5.2 Figured rolled glass

There are many different surface patterns available on figured rolled glass. In general, it can be said that the figured side of the glass should have a clear, sharp pattern free from disfiguring waves.

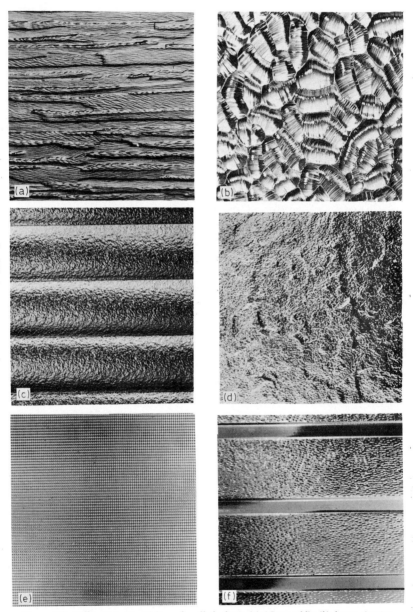

Figure 9.8. Different types of figured, rolled glass; (a) 'Cotswold'; (b) large Arctic; (c) broad reedlyte; (d) clouded cathedral; (e) Pacific glass; (f) pinstripe glass; by courtesy of Pilkington Brothers Limited

88

Vision through the glass is not clear. In general, the light transmission decreases with greater obscurity and diffusion.

9.5.3 Cathedral glass

Cathedral glass is similar to figured, rolled glass. It is usually made in the same way and similar quality requirements can be applied. Cathedral glass as well as figured, rolled glass is also available in colours.

9.5.4 Wired glass

This is a rolled glass having a wire mesh completely embedded in it. A 12 mm ($\frac{1}{2}$ in) square mesh (Georgian wired glass) and a 21·5 mm ($\frac{7}{8}$ in) hexagonal mesh are common types, but other mesh patterns are also available. Wired glass is manufactured in rough cast and polished qualities.

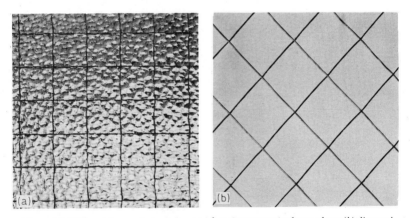

Figure 9.9. Different types of wired glass; (a) $\frac{1}{4}$ in Georgian wired cast glass; (b) diamond mesh polished wired glass; by courtesy of Pilkington Brothers Limited

The mechanical strength of wired glass is only half that of sheet glass (see Section 4.6). When wired glass cracks, however, the embedded wire holds the glass together. Wired glass is usually used as a safety glass against fire. Local regulations should be studied before the glass is installed. The thickness of the glass is usually 5–7 mm ($\frac{3}{16}-\frac{9}{32}$ in).

G

9.5.5 Rolled opal glass

Opal glasses have characteristic light-scattering properties due to inclusion of small particles in the glass. These particles have different indices of refraction from that of the glass itself. The glasses may be white or coloured. Rolled opal glass is available in a rough cast quality as well as polished opal glass. The thickness of the glass is usually between 6 mm and 12 mm ($\frac{1}{4}-\frac{1}{2}$ in).

Opal glass may also be flashed onto a transparent sheet in a thin layer.

9.6 HAND-BLOWN SHEET GLASS

Some sheet glass is still made by hand. This glass has an uneven thickness and contains a relatively high proportion of seeds and blisters. It is usually referred to as antique glass and used mainly for decorative purposes (e.g. in stained glass windows).

9.7 HEAT-ABSORBING GLASSES

As was mentioned in Section 5.5 heat-absorbing glasses absorb a substantial part of the infra-red region of the solar energy and, to some extent, the visible light. Consequently, the temperature of the glass is raised. The heat thus absorbed will be re-radiated outwards and inwards by the glass.

If, for instance, a heat-absorbing glass transmits 42% directly and the reflection is 8%, then the glass will absorb 50% of the heat in the sun's radiation. It can be assumed that one-fifth of the radiated energy will enter the room and four-fifths will pass to the outside. In this particular case the glass will transmit $42 + \frac{1}{5} \times 50 = 52\%$ of the sun's radiation falling on its surface.

Heat-absorbing glasses may be used as single-glazing material. In many places double glazing is preferred, however, and the heat-absorbing glass is then placed on the outside. This gives 10–15% better heat protection than if the heat-absorbing glass is placed on the inside. The heat-absorbing glass may be used in a sealed, double-glazing unit or it may be placed in an auxiliary frame outside the clear window glass. In this latter case it is an advantage to ventilate the space between the two glasses.

It is necessary to allow room for expansion when glazing heat-absorbing glasses. For sheets up to 75 cm (30 in) in length, 3 mm

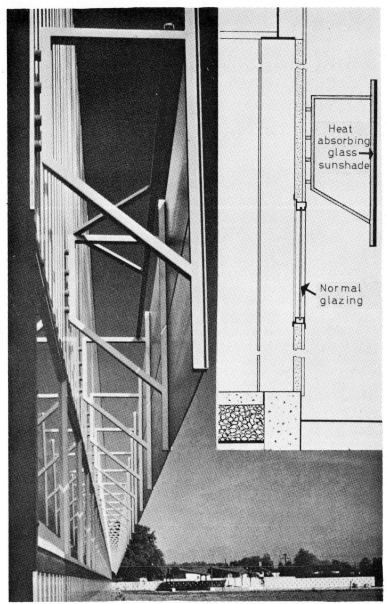

Figure 9.10. Installation of heat-absorbing glass; by courtesy of Pilkington Brothers
Limited

In this factory, to reduce the cooling load, a sunshade spanning the entire western elevation is used. The heat-
absorbing glass is held in a simple frame of steel angles and channels and projected in front and above the clear
glass fixed windows, allowing a free circulation of air between the sunshade and the fixed window.

($\frac{1}{8}$ in) of expansion room will be enough. Larger glass panes will require 5 mm ($\frac{3}{16}$ in) in all four directions.

Some transmittance curves for heat-absorbing glasses are shown in *Figure 5.4*. Thermal data for two of these glasses are given in *Table 9.6*.

Table 9.6. *Approximate thermal data for heat-absorbing glasses and a sealed, double-glazing unit made from ordinary sheet glass. The data are given as a percentage of the solar energy*

Type of glass	Optical characteristics	Primary figures	Secondary figures	
			Passed	Rejected
Grey glass 6·5 mm ($\frac{1}{4}$ in)	Transmittance	40	40	
	Absorptance	52	20	32
	Reflectance	8		8
		100	60	40
Bronze glass 6·5 mm ($\frac{1}{4}$ in)	Transmittance	44	44	
	Absorptance	50	18	32
	Reflectance	6		6
		100	62	38
Double-insulating glass of 6·5 mm ($\frac{1}{4}$ in) bronze glass and a 6·5 mm sheet glass (sheet glass on room side)	Transmittance	37	37	
	Absorptance	55	8	47
	Reflectance	8		8
		100	45	55
Insulating glass with two 6·5 mm ordinary sheet glass panes	Transmittance	73	73	
	Absorptance	17	6	11
	Reflectance	10		10
		100	79	21

It is seen from *Table 9.6* that approximately 80% of the solar energy will pass through double glazing with ordinary sheet glass. About 60% will pass through a single pane of grey glass or bronze glass. When combining a heat-absorbing glass with a sheet glass, less than half of the solar energy will pass through.

9.8 LIGHT- AND HEAT-REFLECTING GLASSES

Light- and heat-reflecting glasses are available in three different types according to the method used in applying the reflecting coating onto the glass. This is explained in Section 5.6.

Some thermal data of different reflecting glasses are given in *Table 9.7*.

Figure 9.11. The figure shows a comparison between a light- and heat-reflecting glass (left above) and ordinary clear glass; the Solarban glass is essentially neutral in colour and provides sufficient light transmission (20 %) by courtesy of Pittsburgh Plate Glass Company

As seen from *Table 9.7* a gold coating on one of the surfaces of a double-glazed unit may stop more than 70% of the solar energy from passing the unit. A nickel-coated, double-glazed unit will stop approximately 60% and so will a single glass with a ceramic, reflecting coating.

Pittsburgh Plate Glass Company have published the following figures for one of their reflecting glasses.

Transmittance:

Visible light	20%
Ultra-violet	11%
Infra-red	2%
Total sun energy	9%

93

Reflectance:

Visible	44%
Total sun energy	46%

The U-value of this glass is as low as 1·7 kcal/m^{2}hdegC (0·35 Btu/ft^{2}hdegF). This is lower than the U-value of a triple glazing unit of sheet glass, which is 1·8 kcal/m^{2}hdegC (0·37 Btu/ft^{2}hdegF). The U-value of the gold-coated, double glazing units is about 1·8 kcal/m^{2}hdegC (0·37 Btu/ft^{2}hdegF) and that of nickel-coated glasses 2·0–2·4 kcal/m^{2}hdegC (0·40–0·50 Btu/ft^{2}hdegF).

Table 9.7. Approximate thermal data for light- and heat-reflecting glasses. The data are given as a percentage of the solar energy

Type of glass	Optical characteristic	Primary figures	Secondary figures	
			Passed	Rejected
Single pane with a ceramic coating	Transmittance Absorptance Reflectance	26 56 18	26 15	41 18
		100	41	59
Double glazing unit with a vacuum-evaporated gold coating on the inner surface of the outer glass pane	Transmittance Absorptance Reflectance	24 36 40	24 4	32 40
		100	28	72
Double glazing unit with a chemically-precipitated nickel coating on the inner surface of the outer glass pane	Transmittance Absorptance Reflectance	25 52 23	25 13	39 23
		100	38	62

9.9 PROCESSED GLASS

An increasing number of finishing operations are used in order to upgrade the flat glass or to create new flat glass products. Some of these operations take place while the glass is hot and soft and others are carried out at room temperature.

Figure 9.12. Solarpane light- and heat-reflecting glass installed at the head office of
AB Volvo, Gothenberg, Sweden; by courtesy of Oxelösund Järnverk

95

9.9.1 Toughened glass

The mechanical strength of glass will increase if compressive stresses are introduced at the glass surface and balancing tensile stresses in the interior. This condition can be obtained by either heat treatment or chemical treatment of the glass.

9.9.1.1 Thermal strengthening—It has long been known that the strength of glass can be increased by heating it to a high temperature and then cooling it rapidly. The glass must be heated to a temperature above its annealing point, usually 650°C (1200°F). It is then cooled rapidly by means of air jets. The surface of the glass will then be in compression, balanced by tension inside. Permanent stresses are thus set up in the glass.

A thermally-tempered glass (a toughened glass) cannot be broken unless sufficient force is applied which will first overcome the compression at the surface and then introduce enough tension for breakage. A crack will always start at the surface and since the surface is in compression in tempered glass, its strength will be much higher

Figure 9.13. Testing of tempered (left) and ordinary flat glass of similar thicknesses; notice how the tempered pane has lifted from the bricks after the steel ball has hit it

than that of annealed glass. Tempered glass is usually three to five times stronger than annealed glass for impact, sustained loads and thermal shock; however, the surface of the glass is not harder than annealed glass.

When thermally tempered glass breaks it fractures into a great many small pieces. These small pieces are interlocking and have no large jagged edges. They are therefore not likely to inflict serious wounds.

Tempered glass cannot be cut. Exact sizes must therefore be specified to the manufacturer. Tempered glass used in motor vehicles is defined in BS 857 and in the American Standard Safety Code for Safety Glazing Materials for Glazing Motor Vehicles Operating on Land Highways (226.1–1950). These standards give methods for testing tempered glass.

Tempered glass is stable up to about 300°C (575°F). It can withstand a thermal gradient of about 300°C (540°F) provided the heat is evenly distributed.

In tempering flat glass the ready-cut panes are usually supported in a vertical position in an electric furnace. Tongs grip the top edge of the pane for support. The temperature of the glass is increased to about 650°C (1200°F) and is then quickly lowered to room temperature by means of air jets. It is very important to support the glass in the correct way and to control the heating and cooling of the glass. Otherwise it is difficult to keep the sheets flat and preserve optically-clear surfaces. The supporting tongs produce small indentations on one edge of the glass. Normal maximum sizes of tempered glass are shown in *Figure 9.14.*

In a new process of tempering glass the panes are kept in a horizontal position in the furnace. The panes ride on gas jets. By this gas-hearth process it is possible to temper thinner glass panes than by the vertical process and still preserve their flatness.

9.9.1.2 Chemically-strengthened glass—Compression at the surface of the glass can also be accomplished by chemical methods. Ion replacement in the surface or surface crystallization may be used in order to get a chemically-strengthened glass. A graphical stress analysis of thermally- and chemically-strengthened glasses is shown in *Figure 9.15.* The surface layer in compression can be made quite thin by the chemical method. This method can give impressive strength figures for the glass. The following figures have been published by Corning Glass Works[1].

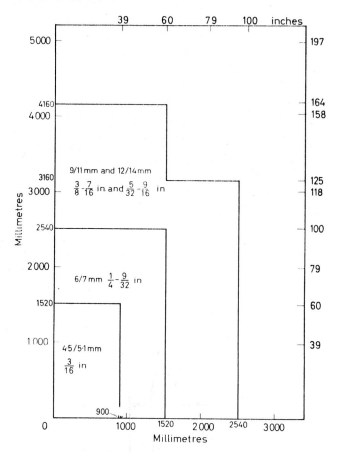

Figure 9.14. Recommended maximum sizes of tempered plate glass of various thicknesses; for tempered sheet the same sizes can be used for 6/7 mm and 4·5/5·1 mm; by courtesy of Pilkington Brothers Limited

Tensile strength of annealed glass—490 kg/cm² (7000 lb/in²)

Tensile strength of thermally-strengthened glass—1400 kg/cm² (20,000 lb/in²)

Tensile strength of chemically-strengthened glass—7000 kg/cm² (100,000 lb/in²)

9.9.2 Laminated glass

Another type of safety glass, besides toughened glass, is laminated glass. Different types of laminated glass are available on the market,

98

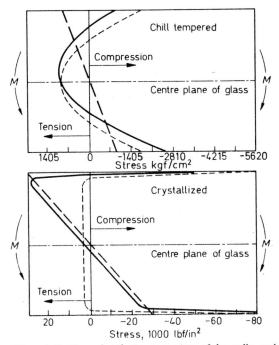

Figure 9.15. Stress distribution in sections of thermally- and chemically-strengthened glass; by courtesy of Corning Glass Works

Two methods for strengthening glass —tempering and one of the techniques in Corning's new Chemcor system —are compared. In both cases, strengthening is achieved by cancelling tensile stress on the surface by pre-stressing in compression. The dashed line is stress created by the external bending load M. Net stress is shown by the solid line. In the chemical strengthening technique, the surface stress in the crystallized glass stays much further into the compression range than in the tempered glass.

but in general, laminated glass is made of two or more panes of glass, firmly united to each other by an interlaying reinforcing material, usually a plastic material (e.g. polyvinyl butyrate). The strength in bending or impact is usually somewhat less than for a solid glass of equal total thickness. When laminated glass fractures, however, the glass fragments will be stuck to the interlaying plastic material. The lamination process is carried out in an autoclave at a pressure of 2–6 kg/cm^2 (28–85 lb/in^2) and a temperature of approximately 100°C (212°F).

Figure 9.16. The largest suspended glass assembly in the world (1966) at Laurel race-course, Maryland, USA; there are 515 panels of $\frac{1}{2}$ in tempered glass and 172 fins, $\frac{3}{4}$ in thick; the total area of glass is 32,000 ft² (3000 m²) and the total weight of glass is 114 tons; by courtesy of Pilkington Brothers Limited

Laminated glass is used for automobile windshields and in many types of architectural glass. A 'bullet-proof' glass (*Figure 9.17*) may be obtained by making a multi-laminated glass. This usually requires a minimum of four layers of glass and three layers of plastic. In special cases it may be required to use toughened glass. Laminated, toughened safety glass may be constructed to give good protection against missiles.

9.9.3 Sealed insulating glass units

Single glazing is often found to give unsatisfactory insulation in windows. Double and triple glazing is therefore used in many places. Prefabricated insulating glass units are available on the market and these are often preferred instead of two or three individual panes of glass in a window.

These prefabricated units consist of two or more panes of glass with cleaned dehydrated air between them. Each unit is hermetically sealed around the edges.

Figure 9.17. Sample of a laminated, bullet-resistant glass that has been hit with a slug from a deadly 45 calibre revolver at a distance of 20 ft (6·1 m); the cartridge, developing 460 ft lb (64 kg) muzzle velocity, shattered the outside layer of glass but was unable to penetrate further; no glass at all has left the reverse surface of the glass; by courtesy of Pittsburgh Plate Glass Company

The insulating glass units available on the market may be divided into three groups according to the type of edge seal used (*Figure 9.18*). *A. Fused glass-to-glass edge seal*—The edges of the glass panes are fused together and this type is therefore an all-glass insulating unit. Dry, clean air is introduced into the space between the two pieces of

101

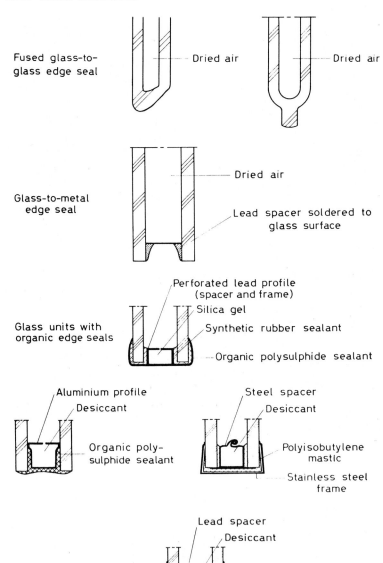

Fused glass-to-glass edge seal — Dried air — Dried air

Glass-to-metal edge seal — Dried air — Lead spacer soldered to glass surface

Glass units with organic edge seals — Perforated lead profile (spacer and frame) — Silica gel — Synthetic rubber sealant — Organic polysulphide sealant

Aluminium profile — Desiccant — Organic poly-sulphide sealant

Steel spacer — Desiccant — Polyisobutylene mastic — Stainless steel frame

Lead spacer — Desiccant — Metal tape — Adhesive — Caulking compound

Figure 9.18. Various types of sealed, insulated glass unit

102

glass before the sealing is completed. Only two glasses can be sealed together using this method of construction and a very satisfactory seal is obtained. The unit is quite rigid.

B. Glass-to-metal edge seal—In this construction a metal spacer is placed between the glass panes. The edge of glass in contact with the spacer is metallized with copper. The spacer is then soldered onto this copper layer. This gives a strong and permanent seal. Before the sealing is completed the air between the panes of glass is exchanged for clean, dry air. An insulating unit of this type may be made up of any number of glass panes. The unit is quite rigid although some flexibility is introduced by the metal spacer.

C. Glass units with organic edge seal—There are many different types of unit in which the seal is made by an organic material. Some units have a glass-to-elastomer and others a glass-to-mastic edge seal. Polysulphide and butyl rubber are often used as sealing materials. The air between the panes of glass is usually dried by means of a desiccant such as silica gel or zeolites of sodium and calcium alumino-silicates (molecular sieves). The desiccant is placed in the metal spacer. There are holes or open slits in the spacer and the desiccant is therefore in permanent contact with the enclosed air.

Due to the organic seal this type of insulating glasses have a mechanical flexibility and are tight and durable. Double- as well as multiple-glazing units can be made. Some units have a metal frame around the glass edges to give better mechanical protection.

The all-glass units of type A are only available as double-glazing units. The air space between the two panes of glass is approximately 6 mm ($\frac{1}{4}$ in). Types B and C can be made with any number of glass panes. Double- and triple-glazing units are most common, however. Units with four panes are used in shop refrigerators. The air spaces as a rule are 6, 9, 12, 15 mm or 18 mm ($\frac{1}{4}$, $\frac{3}{8}$, $\frac{1}{2}$, $\frac{5}{8}$ or $\frac{3}{4}$ in). The most popular distance between the panes of glass is 12 mm ($\frac{1}{2}$ in). Some different types of insulating glass unit are shown in *Figure 9.18*.

The Sealed Insulating Glass Manufacturers' Association of the U.S.A. have issued quality specifications for insulating glass units. According to these specifications the initial dew point of a sealed, insulating glass unit shall be not higher than -51 C ($-60°$F). Certain tests are described by which it is possible to carry out accelerated ageing of the units.

By the use of these SIGMA, or similar tests, it is possible for the manufacturers of sealed, insulating glass units to control the quality

Figure 9.19. Application of the plastics sealant in the manufacture of Polarpane in-sulating glass units; by courtesy of Oxelösunds Järnverk

of their products and to guarantee that they will withstand normal handling and field conditions for a certain period of time.

9.9.4 Cladding glass

Heat-strengthened and ceramic-enamelled glass is often used as a facing material in the cladding of buildings. The glass may be fully or partly toughened. The glass enamel is fused permanently to one of the surfaces of the pane. A wide range of colours is available. It is also possible to carry out additional surface treatments, such as etching and sandblasting, on cladding glasses.

Rough cast, float, sheet and other types of glass may be used for cladding glasses. This type of glass can be used for indoor decoration as well as for the cladding of framed buildings.

Cladding glass will absorb solar radiation and can be quite hot. It is necessary, therefore, to allow certain clearances for expansion. For glasses of a maximum dimension of 80 cm (30 in), 3 mm ($\frac{1}{8}$ in) should be allowed all round. Larger sizes may require 5 mm ($\frac{3}{16}$ in) all round. It is also necessary to allow a certain deviation from complete flatness due to the heat treatment. Cladding glasses are usually 5–10 mm ($\frac{3}{16}$–$\frac{3}{8}$ in) thick, depending on their surface area. Fully heat-

Figure 9.20. Tempered, cladding glass on a building in Washington, D.C.; by courtesy of Pittsburgh Plate Glass Company

105

H

strengthened cladding glass has a mechanical strength equal to that of ordinary tempered glass.

9.9.5 Sandblasting

By blowing a jet of sand or other abrasive, such as carborundum, against the glass a matt or diffusing surface will be obtained. According to the pressure of the jet and the type of abrasive used, the finish will be fine, medium or coarse. Sandblasting gives a decorative effect. Glass for use as chalk boards is often sandblasted on one side and painted on the other.

9.9.6 Acid-etching

A matt surface can also be obtained by acid-etching. The vapour of hydrofluoric acid will readily attack glass, forming silico-fluorides. When glass is dipped in hydrofluoric acid, a nice smooth surface will appear. This is the method used in acid-polishing. The matt surface produced by the vapour of hydrofluoric acid is smoother than that obtained by sandblasting.

A very decorative, frosted surface can be produced by glue-etching. In this process a layer of glue is applied to the glass. After hardening the glue adheres so strongly to the glass that small chips of glass are torn away when the glue is removed from the surface.

9.9.7 Metallizing

Processes for metallizing glass have been known for several hundred years. The Venetians made mirrors in the fourteenth century by coating the glass with an amalgam.

In most current silvering processes, a silver nitrate solution is reduced to metallic silver by reaction with a reducing solution (e.g. glucose). A high quality glass must be used and the glass must be cleaned very carefully before the silvering process.

The glass surface is usually treated with a stannous chloride solution before silvering. The silver solution may be applied by a spray method. The silver coating can then be protected by paint or by electroplated copper and paint.

Metals can also be applied to glass electrolytically. Gold, silver, copper and other metals may be deposited on glass in this way.

Another way of metallizing glass is by means of vacuum distillation and deposition of metals like gold, aluminium, silver and many others. For some special purposes the aluminium may be placed on the front surface of the mirror. Glass can be made to reflect part of the visible light and to transmit another part. This type of glass will act as a mirror when looking at it from a light to a darker room. The transmission is quite good, however, when looking at it from the dark room. Metallized glasses for building purposes are described in Sections 5.6 and 9.8.

9.9.8 Electrically-heated glass

Electrically-heated glass has been used for many years as windshields in aircraft. It has also been used in trains, buildings and cars.

Two types of electrically-heated glass are available. In one type fine metallic wires are used as the heating element. The resistance can be kept quite low and the necessary power can be applied from a 12 V battery.

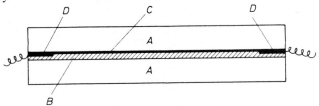

Figure 9.21. Electrically-heated glass. A, glass; B, plastic interlayer; C, conductive coating; D, busses

In the other type of electrically-heated glass, a conductive film is used as the heating medium. One common type of film is stannic oxide, but many other types of conductive coatings can be used. Methods of applying these coatings to the glass surface are described in Section 5.6 and Subsection 9.9.7. The resistivity of different films usually varies between 25 and 100 ohms per square. Since up to 1000 W/m² (93 W/ft²) may be required for a windshield in a car, the power may have to be applied in the 100 V range. Less power is needed for rear windows, however, and a 12 V battery may be adequate.

Electrically-heated glass gives quick, uniform heating and it can be assumed that this type of heating will be increasingly used in homes and automotives in the future.

9.9.9 Bending

When bending flat glass, it is usually placed on a metal or refractory mould. The glass and the mould are placed in a kiln and heated up so that the glass softens and assumes the shape of the mould. The glass is then cooled carefully to give good annealing and to avoid development of any permanent stress. Glass can be bent into many different shapes.

9.9.10 Edge-work and bevelling

It is often necessary to grind the edges of flat glass products. Glass to be used as shelves, for instance, must have rounded edges (*Figure 9.22*). Removal of the sharp cut edge of a glass is usually referred to as arrising. A full treatment of edge-working, however, usually

Figure 9.22. Edge work on sheet glass intended for glass shelves; by courtesy of Oxelösunds Järnverk

Table 9.8. *Some details of edge work and bevelling as given in BS 952:1964*

Title	Form	Finish	Illustration
(1) Arris edge	A small bevel of width not exceeding $\frac{1}{16}$ in, at an angle of approximately 45° to the surface of the glass	Ground, smoothed or polished	$\frac{1}{16}$ in(1·59 mm) or less
(2) Flat edge	The cut edge of the glass is flat and the surface edges are slightly arrised	Ground, smoothed or polished	
(3) Round edge	The cut edge of the glass is slightly curved to form an arc of a circle of selected radius	Ground, smoothed or polished	
(4) Half round	Half of the cut edge of the glass is rounded approximately in the form of a quarter circle. The remaining surface edge is slightly rounded	Ground, smoothed or polished	
(5) Full round	The cut edge of the glass is rounded approximately in the form of a semi-circle	Ground, smoothed or polished	
(6) Thumb or bullnose	The surface of the glass is curved in a shape resembling the profile of a thumb	Ground, smoothed or polished	
(7) Bevel	The surface edge of the glass is bevelled to $\frac{1}{8}$ in or more in width, as required. The angle formed by the intersection of the plane of the bevel with the face of the glass is about $7\frac{1}{2}°$	The bevel is polished unless otherwise specified. The nose of the bevel is left as cut	$7\frac{1}{2}°$ (3·2 mm min) $\frac{1}{8}$ in min.

109

includes grinding, smoothing and polishing. The grinding is usually carried out with sand and the polishing with rouge as in the process for making plate glass.

Grinding and polishing may be carried out by different edge-working machines. Some details of edge-work and bevelling as specified in BS 952: 1964 are given in *Table 9.8*

REFERENCES AND BIBLIOGRAPHY

[1] 'Corning Achievement', *Chem. Engng, Albany*, November 11, 1963
[2] EITEL, W. and PIRANI, M. *Glastechnische Tabellen*, Springer, Berlin, 1932
[3] PHILIPS, C. J. *Glass, its Industrial Applications*, Reinhold, New York, 1960
[4] SHAND, E. B. *Glass Engineering Handbook*, Maple Press, York, Pa., 1958
[5] McGRATH, R. and FROST, A. C. *Glass in Architecture and Decoration*, Architectural Press, London, 1961
[6] VÖLCKERS, O. *Tafelglasdaten*, Verlag Karl Hofmann, Schorndorf bei Stuttgart, 1954
[7] SEIZ, R. *Glaser-Fachbuch*, Verlag Karl Hofmann, Schorndorf bei Stuttgart, 1963
[8] SPICKERMANN, H. *Erweitertes Gussglas Tabellarium*, Verlag Karl Hofmann, Schorndorf bei Stuttgart, 1958
[9] SCHNECK, A. G. *Fenster,* Julius Hofmann Verlag, Stuttgart, 1963
[10] WIGEN, R. 'Vinduer', *Handb. Norg. ByggsforskInst.* 15 (1963)
[11] GJELSVIK, T. 'Tests with Factory-sealed, Double-glazed Window Units', *Rapp. Norg. ByggsforskInst.* 33 (1962)
[12] LINDSKOUG, N. E. 'Views on Electrical Space Heating', *HSB*, Dept. of Building Research, Stockholm, 1964
[13] JEBSEN-MARWEDEL, H. *Tafelglas*, Verlag Girardet, Essen, 1950
[14] PIGANIOL, P. *Les Industries Verrieres*, Dunod, Paris, 1966
[15] PETER, J. *Design with Glass,* Reinhold, New York, 1964
[16] SCHAUPP, W. *Die Aussenwand*, Verlag Georg Callway, Munich, 1962
[17] MARKUS, T. A. *Daylight with Insulation*, Pilkington Bros. Ltd., St. Helens, 1960
[18] BS 952, *Classification of Glass for Glazing and Terminology for Work on Glass,* British Standards Institution, London, 1964
[19] Federal Specification DD-G-451a, *Glass, Flat and Corrugated, for Glazing, Mirrors and other Uses*, Washington, D.C., 1949
[20] PERSSON, R. Wärmeabsorbierende und Wärmereflektierende Gläser', *VDI Z.* Dusseldorf, 1 (1968)

10

GLAZING INSTRUCTIONS

IT is very important to glaze flat glass products properly. Glazing instructions are usually given by the glass manufacturer and these should always be followed. This is especially true when glazing insulating glass units. Most manufacturers give a service guarantee for their insulating glass units but they usually state that the units must be glazed according to their instructions.

10.1 GLAZING OF SINGLE GLASS

The Flat Glass Jobbers Association of Topeka, Kansas, U.S.A. has published a glazing manual[1] which is reproduced here with their kind permission.

Table 10.1. Various edge clearances and tolerances for single lights of glass (Methods 1 to 13)

Glass type and thickness	Glass size		Tolerances		Clearance at head sill and jambs (allow)	Rabbet depth (min.)	Block setting height (range)
	Area (ft^2)	Width or height (in)	Glass cutting size (in)	Sash daylight opening (in)			
Sheet glass							
S.S.	5	40	$\pm\frac{1}{32}$	$\pm\frac{1}{16}$	$\frac{1}{16}$	$\frac{3}{8}$	None req.
S.S.	14	50	$\pm\frac{1}{32}$	$\pm\frac{3}{32}$	$\frac{1}{8}$	$\frac{7}{16}$	$\frac{1}{16} - \frac{3}{16}$
D.S.	5	40	$\pm\frac{1}{32}$	$\pm\frac{1}{16}$	$\frac{1}{16}$	$\frac{3}{8}$	None req.
D.S.	25	80	$\pm\frac{1}{32}$	$\pm\frac{3}{32}$	$\frac{1}{8}$	$\frac{7}{16}$	$\frac{1}{16} - \frac{3}{16}$
$\frac{3}{16}$	25	120	$\pm\frac{1}{16}$	$\pm\frac{3}{32}$	$\frac{11}{64}$	$\frac{1}{2}$	$\frac{3}{32} - \frac{1}{4}$
$\frac{3}{16}$	70	120	$\pm\frac{1}{16}$	$\pm\frac{7}{32}$	$\frac{15}{64}$	$\frac{5}{8}$	$\frac{3}{32} - \frac{3}{8}$
$\frac{7}{32}$	25	120	$\pm\frac{1}{16}$	$\pm\frac{3}{32}$	$\frac{11}{64}$	$\frac{1}{2}$	$\frac{3}{32} - \frac{1}{4}$
$\frac{7}{32}$	70	120	$\pm\frac{1}{16}$	$\pm\frac{7}{32}$	$\frac{15}{64}$	$\frac{5}{8}$	$\frac{3}{32} - \frac{3}{8}$

Table 10.1 *continued*

Polished plate (clear and tinted)

$\frac{1}{8}$	25	128	$\pm\frac{1}{16}$	$\pm\frac{3}{32}$	$\frac{11}{64}$	$\frac{1}{2}$	$\frac{3}{32}\ \frac{1}{4}$
$\frac{1}{8}$	67	128	$\pm\frac{1}{16}$	$\pm\frac{7}{32}$	$\frac{15}{64}$	$\frac{5}{8}$	$\frac{3}{32}\ \frac{3}{8}$
$\frac{1}{4}$	100	120	$\pm\frac{1}{16}$	$\pm\frac{3}{32}$	$\frac{11}{64}$	$\frac{1}{2}$	$\frac{3}{32}\ \frac{1}{4}$
$\frac{1}{4}$	140	156	$\pm\frac{1}{16}$	$\pm\frac{3}{16}$	$\frac{1}{4}$	$\frac{5}{8}$	$\frac{1}{8}\ \frac{1}{8}$
$\frac{1}{4}$	207	220	$\pm\frac{3}{32}$	$\pm\frac{7}{32}$	$\frac{11}{32}$	$\frac{3}{4}$	$\frac{3}{16}\ \frac{1}{2}$
$\frac{5}{16}$	207	220	$\pm\frac{3}{32}$	$\pm\frac{7}{32}$	$\frac{11}{32}$	$\frac{3}{4}$	$\frac{3}{16}\ \frac{1}{2}$
$\frac{3}{8}$	258	286	$\pm\frac{3}{2}$	$\pm\frac{5}{32}$	$\frac{3}{8}$	$\frac{3}{4}$	$\frac{1}{4}\ \frac{1}{2}$
$\frac{3}{8}$	258	286	$\pm\frac{3}{32}$	$\pm\frac{17}{32}$	$\frac{7}{16}$	$\frac{7}{8}$	$\frac{1}{4}\ \frac{5}{8}$
$\frac{1}{2}$	258	286	$\pm\frac{1}{8}$	$\pm\frac{1}{4}$	$\frac{7}{16}$	$\frac{7}{8}$	$\frac{1}{4}\ \frac{5}{8}$
$\frac{3}{4}$	258	286	$\pm\frac{1}{8}$	$\pm\frac{1}{4}$	$\frac{7}{16}$	$\frac{7}{8}$	$\frac{1}{4}\ \frac{5}{8}$
1	76	148	$\pm\frac{5}{32}$	$\pm\frac{7}{32}$	$\frac{7}{16}$	$\frac{7}{8}$	$\frac{1}{4}\ \frac{5}{8}$
1	76	148	$\pm\frac{5}{32}$	$\pm\frac{11}{32}$	$\frac{1}{2}$	1	$\frac{3}{4}\ \frac{3}{4}$
$1\frac{1}{4}$	76	148	$\pm\frac{3}{16}$	$\pm\frac{5}{16}$	$\frac{1}{2}$	1	$\frac{3}{4}\ \frac{3}{4}$

Spandrel glass and heat-strengthened glass

$\frac{1}{4}$	25	80	$+0-\frac{3}{16}$	$+0-\frac{3}{16}$	$\frac{5}{32}$	$\frac{1}{2}$	$\frac{1}{16}\ \frac{1}{4}$
$\frac{1}{4}$	84	168	$+0-\frac{3}{16}$	$+\frac{1}{16}-\frac{1}{4}$	$\frac{1}{4}$	$\frac{5}{8}$	$\frac{3}{8}\ \frac{3}{8}$
$\frac{3}{8}$	25	80	$+0-\frac{3}{16}$	$+0-\frac{3}{16}$	$\frac{5}{32}$	$\frac{1}{2}$	$\frac{1}{16}\ \frac{1}{4}$
$\frac{3}{8}$	84	168	$+0-\frac{3}{16}$	$+\frac{1}{16}-\frac{1}{4}$	$\frac{1}{4}$	$\frac{5}{8}$	$\frac{3}{8}\ \frac{3}{8}$

Figure 10.1 (see opposite page). Glazing details for single glass—Flat Glass Jobbers Association, Topeka, Kansas, USA

A and *B* dimensions shown on chart (see *Table 10.1*).
C dimension $\frac{1}{16}$ in for glass up to 5 ft², $\frac{1}{8}$ in minimum on all over 5 ft², except for method 6 which is zero.
D dimension for methods 1 and 3, minimum $\frac{3}{8}$ in; for methods 5–6 and 11, dimension is zero; other methods, dimension is $\frac{1}{8}$ in.

Typical methods

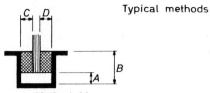

Method 22
Tape glazing
synthetic polymer base

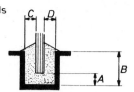

Method 23
Gasket glazing
(neoprene butyl) etc

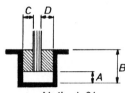

Method 24
Polysulphide compound

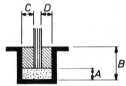

Method 25
Polysulphide with heel
bead elastic glazing
topping both sides

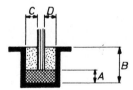

Method 26
Multiple seal polysulphide
heel bead with tape
(synthetic base)

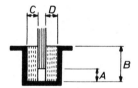

Method 27
Impregnated
urethane foam

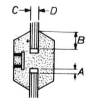

Method 28
Structural rubber
zip-type gasket

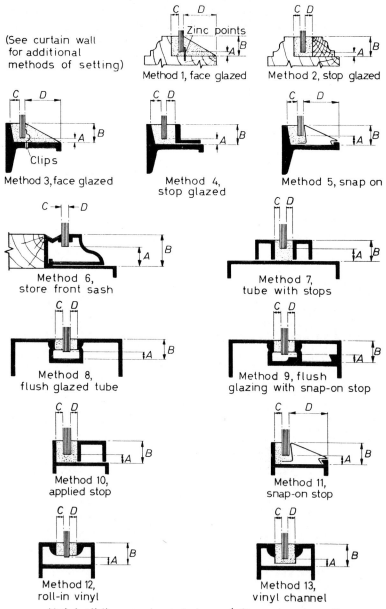

(See curtain wall
for additional
methods of setting)

Method 1, face glazed

Method 2, stop glazed

Method 3, face glazed

Method 4,
stop glazed

Method 5, snap on

Method 6,
store front sash

Method 7,
tube with stops

Method 8,
flush glazed tube

Method 9, flush
glazing with snap-on stop

Method 10,
applied stop

Method 11,
snap-on stop

Method 12,
roll-in vinyl

Method 13,
vinyl channel

Metal sliding sash and doors (often pre-glazed)

Figure 10.2. See opposite page for caption.

Table 10.2. Curtain wall—charts showing clearances

With the exception of setting Method 28; the depth of rabbet and edge clearance—A and B dimensions; C and D dimensions for six methods, single glass (22 to 27) is $\frac{1}{8}$ in

A edge clearance Size (ft²)	$\frac{3}{16}$	$\frac{7}{32}$	$\frac{1}{4}$	$\frac{5}{16}$	$\frac{3}{8}\ \frac{1}{2}\ \frac{3}{4}$	1	C and D Dimension (in)
			Thicknesses (in)				
Up to 25	$\frac{1}{4}$	$\frac{1}{4}$	$\frac{1}{4}$	$\frac{1}{4}$	$\frac{1}{4}$	$\frac{3}{8}$	$\frac{1}{8}$
Over 25 to 70	$\frac{1}{4}$	$\frac{1}{4}$	$\frac{1}{4}$	$\frac{1}{4}$	$\frac{1}{4}$	$\frac{3}{8}$	$\frac{1}{8}$
Over 70 to 84			$\frac{1}{4}$	$\frac{1}{4}$	$\frac{1}{4}$	$\frac{3}{8}$	$\frac{1}{8}$
Over 84			$\frac{1}{4}$	$\frac{1}{4}$	$\frac{9}{32}$	$\frac{1}{2}$	$\frac{1}{8}$
B rabbet depth							
Up to 25	$\frac{5}{8}$	$\frac{5}{8}$	$\frac{5}{8}$	$\frac{5}{8}$	$\frac{5}{8}$	$\frac{3}{4}$	$\frac{1}{8}$
Over 25 to 70	$\frac{3}{4}$	$\frac{3}{4}$	$\frac{3}{4}$	$\frac{3}{4}$	$\frac{3}{4}$	$\frac{7}{8}$	$\frac{1}{8}$
Over 70 to 84			$\frac{3}{4}$	$\frac{3}{4}$	$\frac{7}{8}$	1	$\frac{1}{8}$
Over 84			$\frac{3}{4}$	$\frac{3}{4}$	$\frac{7}{8}$	1	$\frac{1}{8}$

Their instructions for glazing single glass and curtain walls, respectively are given in *Figure 10.1* and *Figure 10.2* and in *Tables 10.1–10.4*.

Figure 10.2. Glazing curtain walls, steel or aluminium; Methods 22 to 28—Flat Glass Jobbers Association, Topeka, Kansas, USA

Consult preparation before glazing, glazing materials and general glazing conditions. The following are some special notes on curtain wall preparation by others.

(1) All openings shall be square, plumb and true in plane.

(2) Framing shall be rigid and all final adjustments made prior to glazing.

(3) All steel shall be primed and dry.

(4) Rabbets and grooves shall be clean (no bolt heads, welds or other obstructions).

(5) The deflection of any metal framing member, in a direction normal to the plane of the wall, shall not exceed $\frac{1}{175}$ of the clear span of the member or $\frac{3}{4}$ in, whichever is least, except that when a plastered surface is affected this deflection shall not be more than $\frac{1}{360}$ of the clear span.

(6) The deflection of any member, in a direction parallel to the wall plane, when the member carries its full design load, shall not exceed 75% of the design clearance dimension between that member and the panel, sash, glass or other part immediately below it.

The glazier shall remove all protective coatings or films from glazing surfaces with solvent from aluminium curtain wall (must not etch or stain aluminium).

See *Table 10.2* for dimensions A and B in curtain wall construction.

These various methods are shown to illustrate some of the current methods. Many new products are coming on the market that may be satisfactory in every respect. Architects should consult with the manufacturers of such products.

Glazing materials used shall be specified by the architect. Consult local building codes for size and thickness requirements.

See *Tables 10.2–10.4* for clearances and glazing materials.

115

Table 10.3. Curtain Wall
Structural rubber zip-type gasket setting Method 28

* B dimension (ft^2)	Glass thickness (in)			
	$\frac{3}{16}$ or $\frac{7}{32}$	$\frac{1}{4}$	$\frac{5}{16}$	$\frac{3}{8}$ $\frac{1}{2}$ or $\frac{3}{4}$
Up to 25	$\frac{1}{2}$	$\frac{1}{2}$	$\frac{1}{2}$	$\frac{5}{8}$
Over 25 to 75	$\frac{1}{2}$	$\frac{1}{2}$	$\frac{5}{8}$	$\frac{5}{8}$
Over 75 to 100		$\frac{5}{8}$	$\frac{5}{8}$	$\frac{3}{4}$

For glass larger or thicker than shown, consult manufacturer.

All A dimensions are $\frac{1}{16}$ in.

All C and D dimensions are zero.

* B dimensions have been developed from current information in manufacturers' literature. The art is so new that complete accuracy of data given is not guaranteed. It is suggested that the architect consults the manufacturer for complete recommendations, as the shape of gasket used may be a factor in deciding the grip on the glass.

Table 10.4: see opposite page.

10.2 GLAZING OF INSULATING GLASS UNITS

Glazing instructions for insulating glass units have been published by the Insulation Glazing Association, London, England[2]. These instructions are given below with kind permission.

Specification of materials
Glazing Compounds

With the exception of Type 1 compound, all glazing compounds must meet the I.G.A. requirements.

Type 1—Metal Casement Putty. Usually applied by hand or knife.

Type 2—Non-setting glazing compound. Usually applied by hand or knife but may be provided in the form of an extruded section. The consistency may be adjusted by the manufacturers for application by gun but the material is still of·a putty-like nature and even when gunned can be handled.

Type 3—A gun applied non-setting compound. Usually single part and non-curing. The material is viscous, tenacious, of semi-fluid consistency and is impossible to apply by hand.

Type 4—An extruded non-setting compound available in a wide range of sections for use without distance pieces and with a capping on the weather face.

116

Table 10.4. Curtain walls;
glazing materials and spacers

Method number	Spacers	Setting blocks 2 required	Glazing materials
22 Tape glazing†	Resilient 40–50 durometer	70–90 durometer	Extruded polybutane tape sealant
23 Gasket glazing	None	70–90 durometer	Vinyl or neoprene gasket 50–70 durometer
24 Polysulphide compound	Resilient 40–50 durometer	70–90 durometer	* Polysulphide compound 20–40* shore *A* hardness
25 Polysulphide with heel bead elastic compound topping	Resilient 40–50 durometer	70–90 durometer	Polysulphide compound 20–40 shore *A* hardness at sill, elastic glazing compound at sides and top*
26 Multiple seal Polysulphide heel bead with tape (synthetic base)	Resilient 40–50 durometer	70–90 durometer	Polyisobutylene butyl or non-oiling polybutene
27 Impregnated urethane foam	None	70–90 durometer	Asphalt or polyisobuty lene impregnated foam
28 Structural rubber zip-type gasket	None	None	Neoprene preformed gasket

* Determined after 7 days ageing at 77°F.

† Resilient tapes of polyisobutylene or butyl may not require spacers.

When elastic glazing compound or other oil-base materials are used with curing-type sealants such as polysulphide-base compounds, always place the polysulphide compound below the oil-base or elastic compound.

Type 5—Elastomeric synthetic rubber sealants which exhibit chemical curing properties, so changing from the viscous semi-fluid state to a cured rubbery mass in the joint. (These should at least comply with the *performance* requirements of British Standard 4254: 1967 for Two-Part Polysulphide–Based Sealing Compounds for the Building Industry.)

117

Setting Blocks

Setting blocks should be of hammered lead, rigid nylon or sealed hardwood such as oak or teak. They should be 1 in to 6 in long (according to the size and type of window) $\frac{1}{8}$ in, or $\frac{3}{16}$ in thick and $\frac{1}{8}$ in wider than the unit thickness when used with Type 2, 3 or 5 Compound and of the width of the unit when used with Type 4 compound.

Location Blocks

Location blocks (of plasticized P.V.C. of a B.S. softness number 35–45, B.S. 2571 : 1963) should be 1 in long for all opening windows except horizontally pivoted windows which are reversible (in this case those on the top edge should be 3 in–6 in according to the size of the unit); they should be of a thickness to suit the edge clearance and $\frac{1}{8}$ in wider than the unit thickness when used with Type 2, 3 or 5 Compound and of the width of the unit when used with Type 4 Compound.

Distance Pieces

Distance pieces should be of plasticized P.V.C. of a B.S. softness number 35–45 (B.S. 2571 : 1963).

They should be 1 in long and of a height to suit the depth of the rebate and the method of glazing; and of a thickness to suit the face clearances.

They are available in the following dimensions:

Height	Thickness
$\frac{1}{4}$, $\frac{3}{8}$ and $\frac{1}{2}$ in	$\frac{3}{32}$, $\frac{1}{8}$, $\frac{5}{32}$, $\frac{3}{16}$, $\frac{7}{32}$ and $\frac{1}{4}$ in
$\frac{5}{8}$, $\frac{3}{4}$ and 1 in	$\frac{1}{8}$, $\frac{3}{16}$ and $\frac{1}{4}$ in

Shims

Shims of $\frac{1}{32}$ in thickness should be of rigid material such as unplasticized P.V.C. in the same sizes as the distance pieces.

Clips, Cleats, Sprigs

They should be sufficient in number and strong enough to withstand the wind pressure or suction to which the unit will be subjected and should be such that they can be covered with at least $\frac{1}{8}$ in of glazing compound.

118

Design considerations

Key to hatching for glazing compounds (applying to all figures):

Type 1   Type 4

Type 2   Type 5

and 3

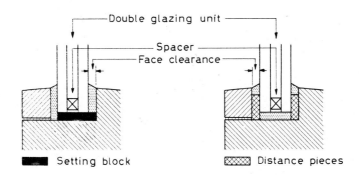

Setting block   Distance pieces

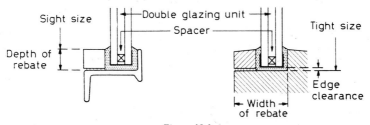

Figure 10.3

Unit Manufacturing Tolerances

Units	Tolerances on Sizes	Tolerances on Thicknesses (for two leaves of glass)		
Up to 30 ft²	$+\frac{1}{8}$ in - 0	Each leaf up to $\frac{1}{4}$ in .. $+\frac{1}{16}$ in $-\frac{1}{32}$ in	$\left.\rule{0pt}{2.2em}\right\}$	On
30–90 ft²	$+\frac{3}{16}$ in - 0	For leaves of $\frac{1}{4}$ in and over .. $+\frac{3}{32}$ in		nominal unit
Over 90 ft²	$+\frac{1}{4}$ in - 0	$-\frac{1}{32}$ in		thickness

From the sight and tight sizes provided it is the responsibility of the unit makers to supply units giving the minimum edge clearances and providing $\frac{3}{8}$ in minimum edge cover.

Frame Tolerances
Metal or Wood $\pm\frac{1}{16}$ in
Tolerances for frames made from other materials such as plastics,

119

concrete, natural or artificial stone will vary and the manufacturer must be consulted.

Recommended Glazing Sizes

All units should be of a size and thickness of glass adequate for the exposure conditions. (See C.P. 152: 1966).

Clearance and Rebates

Face Clearances—Between unit and bead, and unit and rebate, or between unit and rebate when front putty is used:
$\frac{1}{8}$ in minimum for Types 2, 3 and 4 Compounds and for Type 5 Compound used as a capping.
Note: For S.M.W. range of metal windows, it is permissible to reduce the face glazing clearance to $\frac{3}{32}$ in.

Notes

(1) When glazing is into dark coloured frames, or frames which will be finished a dark colour shortly after glazing, the compound manufacturer should be consulted as special compounds may be required.

(2) In all cases the width of the rebate should be sufficient to accommodate the nominal thickness of the unit plus manufacturing tolerances plus face clearance(s) as required and plus either the bead or face putty.

(3) For units with protective metal foil edging deeper rebates may be required.

(4) Frames, beads and fixings shall be of adequate strength to support and restrain the glazed unit and to withstand wind loadings.

(5) When units exceeding $\frac{5}{8}$ in nominal thickness are to be glazed

Edge Clearances and Depths of Rebate			Minimum Depth of Rebate (in)	
Area Up to (ft²)	Nominal Thickness of Unit (in)	Minimum Edge Clearance (in)	For Type 2, 3 and 4 Compounds	When using Type 5 Compound as a capping
30	Up to $\frac{5}{8}$	$\frac{1}{8}$	$\frac{1}{2}$	$\frac{5}{8}$
90	Over $\frac{5}{8}$ and up to 1	$\frac{3}{16}$	$\frac{5}{8}$	$\frac{3}{4}$
Over 90 ft² or over 1 in thick			Obtain the instructions of the compound manufacturer at the design stage	

120

into opening windows, special consideration may need to be given to the design of the frame and to the method of glazing. In all cases consult the frame manufacturer.

Front Putty Glazing

Units glazed with a non-setting compound between the unit and rebate and a fronting of metal casement putty or Type 1 Compound (see *Method 1, page* 124) should not exceed the following sizes:

Wind Loading* Up to (lbf/ft²)	Maximum Size Length plus Breadth (in)
15	96
16–20	84
21–30	72
Over 30	Beads should be used for all sizes

* Three second gust

Glazing with Beads

1. *Metal*—(*a*) Angular, solid or tubular beads, screw fixed (*Figure 10.4*).

(*b*) Channel type beads screw fixed or clipped over studs provided that water is not led to undrained spaces (*Figure 10.5*).

(*c*) Channel type beads inserted into grooves provided that water is not led to undrained spaces (*Figure 10.6*).

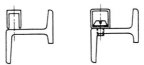

121

ı

(*d*) Face beads set inward or outward as may be necessary to accommodate the thickness of unit and compound (*Figure 10.7*).

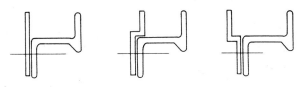

Minimum Dimensions for Front Putty (in)

	Rebate		Nominal Thickness of Unit	Minimum Front Putty	
	Depth	Width		Height	Width
(i) Metal S.M.W.	$\frac{1}{2}$	$\frac{7}{8}$	$\frac{7}{16}$	$\frac{9}{16}$	$\frac{5}{16}$
W.20	$\frac{1}{2}$	$1\frac{1}{8}$	$\frac{9}{16}$	$\frac{9}{16}$	$\frac{7}{16}$
(ii) Wood and Concrete	$\frac{1}{2}$	—	$\frac{9}{16}$	$\frac{9}{16}$	$\frac{3}{8}$

2. *Wood*—The dimensions, cross-section and quality of timber of the bead should be such that shrinkage, warping, flexing between the fixing points and lifting at the corner will not occur. The screw fixing points shall be provided with cups.

3. *Fixing Points*—The fixing points of metal beads held with screws, studs or clips must be located not less than 3 in from each corner and spaced not more than 9 in apart. Timber beads must be held with screws located not less than 2 in from each corner and spaced not more than 9 in apart.

Tolerances: To maintain face clearances the tolerance of bead fixing hole positions shall be $\pm\frac{1}{64}$ in.

Priming, Sealing and Pre-treatment of Rebates or Beads

1. *Wood*—Normally glazing is into softwood which is primed or hardwood which is unprimed. In both cases two coats of sealer supplied by the manufacturer of the glazing compound should be applied to the rebate and inside faces of timber beads to prevent any loss of oil from the glazing compound into the timber. This may only be omitted where the timber surfaces are finished to a gloss finish using a normal non-bituminous paint system. Where it is known or

obvious that timber frames are to be treated with a water repellent or preservative, the compound maker and glazing contractor should be advised so that suitable glazing materials and methods are used. Special requirements obtain where a Type 5 Compound is used (see *pages* 125 *and* 126).

2. *Metal*—Normally glazing is into unprimed or unpainted rebates. Where metal frames are primed with calcium plumbate either in the factory, or on site, it is important that instructions for glazing should not be given unless the priming is sufficiently hard to prevent reaction with the glazing compound.

Site Control

Handling, Storage and Fixing of Frames—All engaged in design and in site control should be concerned with proper discipline and site management to avoid damage to both wood and metal frames, and should ensure the provision of adequate storage to prevent undue moisture absorption by timber frames.

Attention is drawn to the publication of the Metal Window Association Limited entitled *Fixing and Handling on Site* which gives details and illustrations of the procedures which should be adopted for metal windows.

Choice of Glazing Methods

In choosing a method of glazing, consideration should be given to :

(*a*) Maintenance, or frequency of maintenance required for each method. This should be related to the life expectation of the building—commonly 60 years.

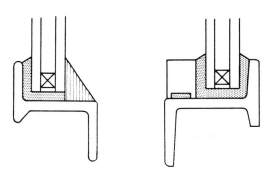

Figure 10.8   *Figure 10.9*

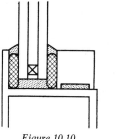
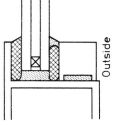

Figure 10.10   Figure 10.11

Glazing Methods

Method 1—Bedding in Type 2 Compound and Fronting with Type 1 Compound. (Type 1 Compound should be adequately protected with paint according to the compound maker's instructions.)

Maintenance Check—Every three years. (*Figure 10.8*)

Method 2—Glazing with Type 2 or Type 3 Compound (with beads).

Maintenance Check—Every five years. (*Figure 10.9*)

Method 3—Glazing with Type 4 Compound, in conjunction with gun or knife forms of Type 2 and Type 3 Compounds as a capping.

(*i*) When compression can be provided by hand pressure—only possible with round, oval or barrel shaped sections of Type 4 Compound. (Some types of sprung on bead do not permit this method of glazing.)

(*ii*) When compression is provided by beads designed to apply pressure mechanically, round, oval, barrel or flat sections of Type 4 Compound may be used.

Maintenance Check—Every five years. (*Figures 10.10* and *10.11*)

Method 4—Glazing with Type 4 Compound in conjunction with Type 2 or Type 3 Compound with a Capping of Type 5 Compound.

(*b*) The availability of permanent access to the weather side or the cost of temporary equipment required, such as scaffolding, cradles, etc.

(*i*) When compression can be provided by hand pressure—only possible with round, oval or barrel shaped sections of Type 4 Compound. (Some types of sprung on beads do not permit this method of glazing.)

(*ii*) When compression is provided by beads designed to apply pressure mechanically, round, oval, barrel or flat sections of Type 4 Compound may be used.

Maintenance Check Every ten years. (*Figures 10.12, 10.13* and *10.14*)

Method 5—Glazing with Type 5 Compound. This method can offer better protection when the unit is to be subjected to severe vibration and shock or where the type of frame design or construction is unsuitable for the adoption of other methods. In all cases the unit manufacturer should be consulted.

Maintenance Check—Every twelve years. (*Figure 10.15*)

Method 6—Glazing with Type 5 Compound in conjunction with Type 2, 3 or 4 Compound. This method is known as the 'heel bead' system. It may be used to ensure that water penetration does not occur underneath the bead and where frames have mitred or butt joints. It is not intended that the top capping should be the main seal. As no visual inspection can be made of the condition of the Type 5 Compound, the only positive method of testing is by means of a water test. In all cases the unit manufacturer should be consulted.

Maintenance Check—Every seven years. (*Figures 10.16* and *10.16A*)

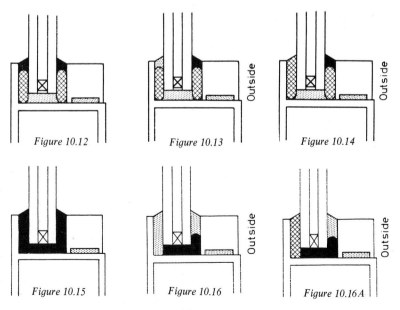

Figure 10.12 *Figure 10.13* Outside *Figure 10.14* Outside

Figure 10.15 *Figure 10.16* Outside *Figure 10.16A* Outside

125

Method 7—Insert Frames. Metal sections of channel formation which are filled with compound and applied to the unit. When joined at the corners they form a completely framed unit which is bedded and securely fixed to the metal window.

Maintenance Check—Where the unit is glazed into the insert frame by one of the preceding glazing methods, the appropriate maintenance period applies.

Method 8—Gaskets.

These are dealt with in I.G.A. Publication No. 3, *Glazing Requirements and Procedures for Compression Gasket Glazing.*

Method 9—Neoprene, P.V.C. or similar insert sections.

At present various designs of sections are used, mainly for single glazing, to prevent water penetration. Used each side of the glass, the second section is either applied with the bead, being retained by grooves, or inserted after the bead is fixed. As there is insufficient information or experience of the application of this method to double glazing units available, there should be consultation between the manufacturers of the unit, frame and the insert sections.

Method 10—Patent Glazing and Dry Glazing Methods.

Systems which are designed to drain water away and not allow it to remain in continuous contact with the edge of the unit may be acceptable.

Insulation between metal and the unit should be provided by lead or P.V.C. blocks between the shoes of patent glazing bars and the edges of the unit. For vertical glazing a continuous sealing strip approved by the unit manufacturer should be used at the bottom to avoid collection of water or the formation of ice under the edge of the units.

In all cases the unit manufacturer should be consulted.

Method 11—Special Applications.

Where unusual conditions such as high humidity or chemical fumes occur as in swimming baths, laundries, chemical factories, high sterility units in hospitals, food factories, etc., and demand special treatment, it will generally be found that Method 4, or Method 5 will give satisfactory results with a suitable Type 5 Compound applied on the face exposed to the unusual conditions. The Type 5 Compound may have to be specially formulated to meet these conditions and the compound maker should be consulted in all cases.

Where conditions of extreme temperature, vibration or other special conditions apply, then the compound maker and frame maker should be consulted at design stage in order that a suitable system is selected.

Where the glazing operation has to be carried out under unusual conditions such as dampness, extremes of temperature or the presence of chemical fumes, then the compound maker must be consulted in order that suitable compounds are selected.

Where access to only one side is available, the unit maker and compound maker must together agree a suitable glazing method.

Note: The maintenance check periods given for Methods 4, 5 and 6 are based on experience with two-part polysulphide sealants conforming to British Standard 4254: 1967 for Two-Part Polysulphide-Based Sealing Compounds for the Building Industry or equivalent standards. (Sealants of different chemical type may also be found to give equivalent performance.)

Inspection

Experience has shown that during the early life of the building certain factors such as settlement, building movement transmitted to the frame or grid, high moisture movement, temperature change due to change from unglazed to glazed conditions, etc., may exert strain or cause damage to the glazing seal. It is important to carry out an inspection, preferably while scaffold or contractors' access equipment is still in place but as late as possible prior to hand over. This inspection is the responsibility of the architect, clerk of works, site agent or general foreman and not of the subcontractor.

The effect of these factors is shown below under *Remedial Work* in the Schedule of Maintenance together with recommended immediate action as a repair to maintain the required performance.

Glazing requirements

1. General

(*a*) Glazing including maintenance and remedial work must be undertaken only by craftsmen trained in the application of the compounds to be used. Adequate control and supervision must be given.

(*b*) Glazing compounds must be able to fulfil the I.G.A. requirements as outlined in this document.

(*c*) Compounds must not be used which have an adverse effect on the edge seal of the unit, the joint of the frame or, if more than one compound is used, upon each other.

(*d*) Order forms should give both sight and tight sizes (see *Measuring requirements.*)

(*e*) The thickness of glazing compound between glass and rebate, and between glass and bead must not be less than that specified under *Design Considerations, Clearances and Rebates* (page 120).

(*f*) *Setting and Location Blocks*

Setting Blocks—Setting blocks which should be used only at the bottom edge of the unit should be of a thickness to locate the unit centrally in the frame. For fixed windows they should be positioned as near quarter points as possible, but where it is necessary to avoid undue deflection of the frame they may be placed nearer to the sides but never less than 3 in from the corner. For vertically pivoted windows they should never be less than 6 in in length.

Location Blocks—Location blocks should be used at the top and sides.

For opening windows, the position of setting and location blocks should be as shown below (*Figure 10.17*).

(*g*) Except for Glazing Methods 8, 9 and 10, it is essential to fill the space between the edges of the unit and the frame with compound to ensure there are no voids before fixing the beads. All beads must be bedded in compound to the frame and unit.

(*h*) The glazing compound, both outside and inside, must be chamfered to prevent water lodgment.

(*i*) *Distance Pieces*—Distance pieces should be used in all cases except where a load bearing strip compound is used.

In the case of Glazing Method 1 (front putty glazing) distance pieces are used only between the unit and rebate. In all other cases they should be used on both sides of the glass opposite each other and spaced as follows:

For beads which are fixed by screws or over studs.

Distance pieces should coincide with the fixing points provided.

For beads which fit into continuous grooves.

The first distance piece should be at approximately 2 in from each corner, and the remainder be located at approximately 12 in centres.

In no case should a distance piece coincide with a setting or location block.

The height of distance pieces when Type 2 and 3 Compounds are used should be $\frac{1}{8}$ in less than the rebate depth: where Type 5 Compound is used as a capping the height of the distance piece should be $\frac{1}{4}$ in less than the rebate depth (*Figure 10.18*).

128

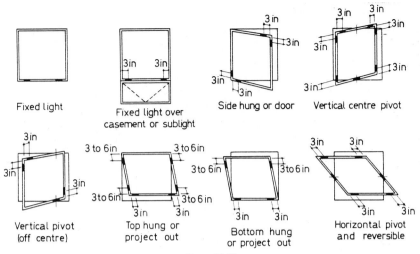

Figure 10.17

2. Measuring

When measuring, take both sight and tight sizes and CHECK that:

(*a*) Any projection such as those caused by beads or bead retaining grooves are allowed for in the tight size.

(*b*) The distance between the face of the rebate and the back of the bead is not less than the nominal thickness of the unit, plus the manufacturing tolerance, and plus the clearances required for the compound.

(*c*) The rebate is deep enough to provide the edge clearance and cover required (*see Design Considerations, Clearances and Rebates,* page 120).

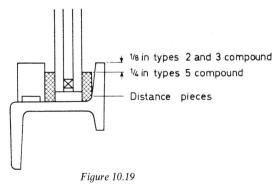

Figure 10.19

129

Site procedure for the glazier

1. Handling and Storage Requirements

All units must be handled with care. Units delivered in cases must be unpacked on arrival. Should any units be found to be wet, they must be dried. All units must be stored on edge on racks in dry conditions. Units must never be laid flat. Supporting blocks of wood or felt should be used to prevent edge damage. The blocks should be set to form a right angle with the back support so that adequate support is given to both panes.

2. General

(*a*) Where units have metal or plastic protective edging which overlaps the surface of the unit, check that this is not damaged. If damaged refer to the unit manufacturer.

(*b*) Check each unit in the opening for edge clearance required.

(*c*) No attempt must be made to alter the size of the unit by nipping or other means.

(*d*) Where units have protective metal foil edging which projects above the sight line, check procedure with the manufacturer.

(*e*) The unit should not be glazed without prior consultation with those concerned if, due to tolerance in the manufacture of the unit or frame or because either is out of square or in wind, the edge or face clearance is reduced to less than that specified elsewhere in this chapter (page 120).

(*f*) Where units have been marked by the manufacturers indicating which should be the top, then this instruction should be followed.

3. Preparation of Rebates and Beads

(*a*) All rebates and beads must be brushed free of dust and be dry.

(*b*) For Type 1, 2, 3 and 4 Compounds—Check that rebates and beads of wood frames have been sealed as described on page 122. If sealing has not been carried out, then apply two coats of sealer supplied for this purpose by the compound manufacturer.

Where it is known or obvious that timber frames have been treated with a water repellent preservative, the recommendations of the compound manufacturer should be obtained.

Rebates in concrete should be sealed with two coats of sealer supplied for this purpose by the compound manufacturer.

(*c*) For Type 5 Compound—When Type 5 Compounds are used,

rebates, beads and glass must in order to obtain effective adhesion, be prepared as follows:

Metal—Where it is necessary to remove any protective coating, such as lacquer, wax, priming or paint, follow the frame and compound manufacturers' instructions.

Wood—Special primers may be needed, especially on teak. The recommendations of the compound manufacturers must be sought and followed.

(*d*) Cleaning of Rebates, Beads and Unit—When Type 5 Compounds are used, cleaning must be carried out as follows, in order to obtain effective adhesion:

(*i*) *Rebates and Beads*

    (*a*) Before glazing clean by means of a degreasing solvent and finish with a further cleaning to remove any residue.

    (*b*) Immediately before applying the capping compound give a second cleaning with a clean dry cloth.

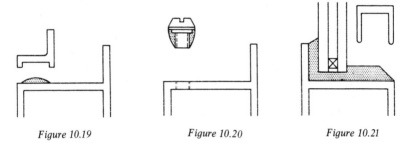

Figure 10.19   Figure 10.20   Figure 10.21

    Note: When cleaning frames, reference should be made to the frame manufacturer because of the possible damaging effect of some cleaning solvents on the frame and on the joint sealant.

(*ii*) *The Unit*

Before glazing, all edges should be wiped with a degreasing solvent, and then be given a final cleaning to remove any residue.

Note: If units have a protective metal or plastic edging, consult the manufacturer on the material to be used for cleaning.

(*e*) *Inspection of Frame Conditions*

(*i*) Metal frames which have been primed or painted should be checked to ensure that the film has dried hard. The primer or paint surface should be rubbed lightly with a clean cloth moistened with white spirit and any removal of the film will indicate that the primer or paint is not sufficiently dry to allow glazing to proceed.

For checking timber frames for wetness a sharp object such as a nail file or 6 in nail should be pressed on to the clean and dried surface of the bottom of the rebate to observe whether water is squeezed out of the timber.

(*ii*) Check that frame and beads are not damaged and that joints in the frame are not open.

If either of these examinations indicates that the frames are not in suitable condition for glazing, the fact should be reported to those concerned, requesting a decision on whether to proceed with glazing.

(*f*) All holes in the rebate, except weep holes for dry glazing, should be filled with compound if they cannot be filled during bedding and glazing.

(*g*) Preparation of Rebates to receive the Beads. The following procedures, which vary according to the type of bead and method of fixing, should be carried out after the unit is in position and all voids filled.

(*i*) For angular, solid or tubular metal beads, and for timber beads:
Apply with gun Type C Compound to bottom of the rebate along the line of the fixing points (*Figure 10.19.*)

(*ii*) For channel type beads fixed over studs: Dip the screw thread of the stud in Type 3 Compound and fix the stud to the frame. Apply Type 2 or Type 3 Compound, whichever is being used for the glazing, to the rebate (*Figures 10.20 and 10.21.*)

(*iii*) For channel type beads inserted into grooves:
Apply Type 2 or Type 3 Compound, whichever is being used for the glazing, to the rebate (*Figures 10.20 and 10.21*).

4. *Glazing Procedures*

Method 1—Bedding in Type 2 Compound and Fronting with Type 1 Compound.

(*a*) Apply sufficient Type 2 Compound to rebate to ensure that when the unit is pressed home there will be a solid bed between the unit and rebate and surplus compound will be squeezed out.

(*b*) Press on to the rebate the appropriate number of distance pieces, of the correct dimensions, at the required positions.

(*c*) Press setting blocks, $\frac{1}{8}$ in wider than the thickness of the unit, hard on to the rebate at the required positions.

132

(*d*) Position the unit in the frame and adjust for clearance. Insert location blocks, $\frac{1}{8}$ in wider than the thickness of the unit, as required for opening windows, and press the unit firmly on to the distance pieces all the way round.

Figure 10.22

(*e*) Fill completely the spaces between the edge of the unit and the frame, eliminating all voids.

(*f*) Trim the compound level and flush with the face of the unit. Fix clips, cleats or sprigs. Clean glass and rebate and then front with metal casement putty (Type 1 Compound) to finish not less than $\frac{9}{16}$ in in height and to the edge of the rebate (*Figure 10.23.*)

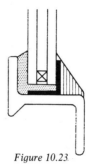

Figure 10.23.

(*g*) Trim off surplus Type 2 Compound at an angle to shed water.

Method 2—Glazing with Type 2 or Type 3 Compound (with beads).

(*a*) Apply sufficient compound to rebate to ensure that when the unit is pressed home there will be a solid bed between the unit and rebate and that surplus compound will be squeezed out.

(*b*) Press on to the rebate the appropriate number of distance pieces, of the correct dimensions, at the required positions.

(*c*) Press setting blocks, $\frac{1}{8}$ in wider than the thickness of the unit, hard on to the rebate at the required positions.

(*d*) Position the unit in the frame and adjust for clearance. Insert location blocks, $\frac{1}{8}$ in wider than the thickness of the unit, as required for opening windows, and press the unit firmly on to the distance pieces all the way round.

133

(*e*) Fill completely the spaces between the edge of the unit and the frame, eliminating all voids.

(*f*) Prepare rebate to receive the beads as described in *Site Procedure*, paragraph 3.

(*g*) Apply compound to face of bead, press in distance pieces of the correct dimensions, in a position to coincide with the distance pieces on the opposite face of the unit (*Figure 10.24.*)

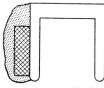

Figure 10.24

(*h*) Apply bead to unit and rebate, and secure to the frame.

(*i*) Check that distance pieces are pressed down to the bottom of the rebate. If distance pieces are not under compression insert shims between distance pieces and bead to obtain compression.

(*j*) Fill in with compound where necessary and trim off to provide a smooth chamfer.

Method 3—Glazing with Type 4 Compound, in conjunction with gun or knife forms of Type 2 and Type 3 Compounds as a capping.

(*i*) When compression can be provided by hand pressure—only possible with round, oval or barrel shaped sections of Type 4 Compound. (Some types of sprung on bead do not permit this method of glazing.)

(*ii*) When compression is provided by beads designed to apply pressure mechanically, round, oval, barrel or flat sections of Type 4 Compound may be used.

Procedure (*When capped both sides*)

(*a*) Apply to the rebate strip compound of adequate thickness and of a depth to finish at least $\frac{1}{8}$ in from the top of the rebate, when it is in contact with the bottom of the rebate.

(*b*) Place setting blocks, the width of the thickness of the unit, hard on to the strip compound, at the required positions.

(*c*) Position the unit in the frame and adjust for clearance. For opening windows insert location blocks, the width of the thickness of the unit, as required. Press the unit firmly on to the strip compound all the way round.

(*d*) Gun or knife compound between the edges of the unit and the frame, completely fill all voids.

(*e*) Prepare rebates to receive the beads as described in *Site Procedure*, paragraph 3.

(*f*) Apply strip compound of adequate dimensions to the unit or bead to finish at least $\frac{1}{8}$ in from the top of the rebate.

(*g*) *For Method 3* (*i*) *above*:

Press beads firmly home, compressing the strip compound, and secure the beads to the frame.

For Method 3 (*ii*) *above*:

Apply beads to the unit and frame. Fix, applying the specified pressure in the manner provided for by the frame manufacturer.

(*h*) Apply compound to fill the spaces between rebate, bead and unit. If applied by knife, trim off to provide a smooth chamfer. If gun applied, it should be so controlled that trimming is not necessary.

Note—(When capped on the weather side only) With both these procedures, when capping is required only on the outside, the inner strip should finish flush with the top of the rebate or bead.

Method 4—Glazing with Type 4 Compound in conjunction with Type 2 or Type 3 Compound with a Capping of Type 5 Compound.

(*i*) When compression can be provided by hand pressure—only possible with round, oval or barrel shaped sections of Type 4 Compound. (Some types of sprung on beads do not permit this method of glazing.)

(*ii*) When compression is provided by beads designed to apply pressure mechanically, round oval, barrel or flat sections of Type 4 Compound may be used.

Procedure (*When capped both sides*)

(*a*) Apply to the rebate strip compound of adequate thickness and of a depth to finish at least $\frac{1}{4}$ in from the top of the rebate, when it is in contact with the bottom of the rebate.

(*b*) Place setting blocks, the width of the thickness of the unit, hard on to the strip compound, at the required positions.

(*c*) Clean glass and rebates—see *Site Procedure*, paragraph 3(*d*).

(*d*) Position the unit in the frame and adjust for clearance. For opening windows insert location blocks, the width of the thickness of the unit, as required. Press the unit firmly on to the strip compound all the way round.

(*e*) Gun or knife compound between the edges of the unit and the frame, completely filling all voids.

(*f*) Prepare rebates to receive the beads as described in *Site Procedure*, paragraph 3(*g*).

(*g*) Clean beads, and apply to the unit or bead strip compound of adequate thickness and of a depth to finish at least $\frac{1}{4}$ in from the top of the rebate.

(*h*) *For Method 4* (*i*) *above*:

Press beads firmly home compressing the strip compound and secure the beads to the frame.

For Method 4 (*ii*) *above*:

Apply beads to the unit and frame. Fix, applying the specified pressure in the manner provided for by the frame manufacturer.

(*i*) Clean the bead, the frame and the glass at the top of the rebate. Prime as required by the compound manufacturer—see *Site Procedure*, paragraph 3(*g*).

Apply Type 5 Compound with a gun, selecting a nozzle of suitable shape and dimensions to ensure that a chamfer finish is provided.

Note: Where it is not necessary to provide a capping of Type 5 Compound on the inside, the capping may be omitted, in which case the Type 4 strip may finish flush with the rebate or bead. Where the internal capping is of Type 2 or Type 3 Compound, the strip should finish $\frac{1}{8}$ in below the top of the rebate or bead.

Method 5—Glazing with Type 5 Compound.

Procedure

(*a*) Clean glass, rebate and beads—see *Site Procedure*, paragraph 3(*d*).

(*b*) Apply sufficient compound to rebate to ensure that when the unit is pressed home there will be a solid bed between the unit and rebate.

(*c*) Press on to the rebate the appropriate number of distance pieces, of the correct dimensions, at the required positions.

(*d*) Press setting blocks, $\frac{1}{8}$ in wider than the thickness of the unit, hard on to the rebate at the required positions.

(*e*) Position the unit in the frame and adjust for clearance. Insert location blocks, $\frac{1}{8}$ in wider than the thickness of the unit, as required for opening windows, and press the unit firmly on to the distance pieces all the way round.

(*f*) Fill completely the spaces between the edge of the unit and the frame, eliminating all voids.

(*g*) Apply compound to rebate and unit, and press in distance pieces of the correct dimensions, in a position to coincide with the distance pieces on the opposite face of the unit.

(*h*) Apply bead to the unit and rebate and secure to the frame.

(*i*) Check that distance pieces are pressed down to bottom of the rebate. If distance pieces are not under compression, insert shims to obtain compression.

(*j*) Apply compound to fill the spaces between bead, rebate and the unit, selecting a nozzle of suitable shape and dimension to ensure that a smooth chamfer is provided.

Method 6—Glazing with Type 5 Compound in conjunction with Type 2, 3 or 4 Compound.

Proceed generally as for Methods 2 and 3, until the unit, setting and location blocks are in position, ensuring as far as it is possible that little or no compound is between the edges of the unit and the frame, then:

(*a*) Clean rebate, bead and glass at the bottom and prime as required by the compound manufacturer.

(*b*) Apply Type 5 Compound, completely filling spaces between the edges of the unit and frame, and apply sufficient to the bottom of the unit and rebate to ensure that approximately half the depth of the rebate will have a solid bed when the beads are applied.

(*c*) Press in distance pieces of the correct dimensions in a position to coincide with other distance pieces on the opposite face of the unit.

(*d*) Apply beads and secure to the frame.

(*e*) Fill in spaces between unit and bead on the weather face, and on the inside when Type 4 strip compound is used. Apply compound on the inside when Type 2 or 3 Compound is used. Trim off to provide a smooth chamfer.

Method 7—Insert Frames (see *page* 126).

Procedure

The method of assembly, glazing and fixing should be in accordance with the manufacturers' instructions.

Methods 8, 9, 10 and 11 (see *page* 126)—Procedures for these methods are not included as in all cases special design considerations have to be taken into account. The instructions of the manufacturers concerned should be followed.

Schedule of Maintenance

This applies where:

(*i*) Initial glazing has been correctly carried out with suitable materials in accordance with I.G.A. recommendations.

K

(*ii*) The wind pressures have not exceeded those allowed for in the design.

(*iii*) Building movement transmitted to the frame or grid has not caused damage to the glazing seal.

Type 1 Compound—(a) Where putty is sound. Lightly sandpaper (taking care to avoid damaging the glass), then dust off and apply one coat of undercoat and at least one coat of gloss paint.

(*b*) *Where there are surface cracks.* Lightly sandpaper the surface (taking care to avoid damaging the glass) fill and point cracks with suitable filler.

When dry, undercoat and paint.

Type 2 and Type 3 Compounds—Where there are surface cracks, or slight extrusion or retraction:

(*a*) *Where used for complete glazing.* Rake out to a depth of $\frac{1}{8}$ in and replace with the same brand of compound as originally used.

(*b*) *Where used as a capping for Type 4 Compound.* Strip off excess compound. Check that bead provides required compression; if not, proceed as for *Remedial Work* (see below).

Type 4 Compound—Where there are slight surface cracks and only slight extrusion or retraction of the compound: Strip off excess compound. Check that the bead maintains required compression, if not, proceed as for *Remedial Work.*

Type 5 Compound—Where there is only slight extrusion or retraction, and no loss of adhesion or cohesive cracking, no maintenance action is required; *but if there is a slight loss of adhesion or cohesive cracking*: Cut out capping to a depth of $\frac{1}{4}$ in, clean and replace with compound of the same brand as originally used.

Remedial Work

Before the start of remedial work involving dismantling of any component of the system (e.g. glazing bead, insert frame) the unit manufacturer's consent should be obtained.

Where initial glazing did not follow I.G.A. recommendations, in that unsuitable compounds or incorrect methods were used or where excessive building movement, excessive wind load, or neglected maintenance has resulted in glazing compound failure or deterioration it may be possible, provided that the unit seal is undamaged, for remedial work to be carried out.

General—Where beads were not bedded to the frame, these must be removed and bedded.

Type 1 Compound—Severe cracking in the depth due to: (*a*) Failure to paint the compound or to paint within the specified time; and/or

(*b*) Inadequate glazing clips, sprigs or cleats, but with adequate setting blocks, distance pieces and location blocks.

Cut out front putty, fit clips, sprigs or cleats as required and refront with the same brand of putty if it meets the I.G.A. requirements.

Type 2 and Type 3 Compounds—(*a*) *When used in conjunction with Type 1 Compound*: Loss of adhesion or severe cracking in depth due to lack of or inadequate glazing clips, sprigs, cleats or distance pieces. *Reglaze in accordance with I.G.A. recommendations.*

(*b*) *When used alone or in conjunction with Type 4 or Type 5 Compound*: Loss of adhesion or severe cracking or extrusion or retraction due to inadequate or lack of distance pieces.

Reglaze in accordance with I.G.A. recommendations.

*Type 4 Compound—*Severe cohesive cracking, loss of adhesion or extrusion or retraction of compound due to excessive wind load. *Reglaze in accordance with I.G.A. recommendations.*

Type 5 Compound—(*a*) *Where used for complete glazing*: Failure of seal due to cohesive cracking, loss of adhesion, extrusion or retraction of compound.

Reglaze in accordance with I.G.A. recommendations.

(*b*) *Where used as a capping*: Severe cohesive cracking or loss of adhesion.

Reglaze in accordance with I.G.A. recommendations.

Stepped Units

*Application—*Stepped units can be used where the depth or width of the rebate is inadequate for the glazing of standard units.

*Glazing Procedure—*Normal glazing as for single glass conforming to C.P. 152, 1966, will apply with the following precautions:

(*i*) The inset glass must always be to the inside.

(*ii*) No attempt must be made to alter the size by nipping or cutting.

(*iii*) The correct edge which is marked must be placed at the bottom.

(*iv*) Setting blocks must be used to provide a clearance between the larger glass and bottom of the rebate, which should also maintain a clearance between the smaller glass and the top of the back of the rebate.

(*v*) The inset glass must not foul fittings such as handles, hinges or locking devices. To avoid this, it may be necessary to insert distance pieces to reduce the projection of the units.

(*vi*) The space between the inset glass and the top of the rebate should be filled with the same type of putty as is used for the glazing.

Where the frame projects beyond the unit, the putty should be finished with a chamfer. Where the unit projects beyond the frame, the putty should finish with a triangular fillet from the edge of the unit to the top of the frame (*Figures 10.25 and 10.26*).

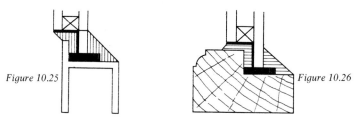

Figure 10.25   Figure 10.26

All-Glass Units

Normally all glass units can be glazed by any of the methods described in this document. However, as they are not supplied with a regular rectangular edge profile, special precautions may sometimes be necessary to ensure that the unit is held firmly and securely in the frame against wind pressure and/or suction. The sealing (pore) hole should be at the top.

Where the rebate depth does not enable rectangular distance pieces to bear on not less than $\frac{1}{8}$ in on both faces of the unit and be covered to the required depth by glazing compound, and glazing is to be carried out using Types 2, 3 or 5 Compounds, then a combined setting block/distance piece of the same profile as that of the edge of the unit, may be required. Instructions from the unit manufacturers should be obtained, preferably at the design stage.

Again where the rebate is as described above and glazing is to be carried out using Type 4 Compound this should be extruded into such a profile that when fixed the unit is held firmly in the opening against wind pressure or suction. The compound should thus make contact with the faces of the unit, the frame and the bead, and fill the space between all components completely.

REFERENCES

[1] *Glazing Manual*, Flat Glass Jobbers Association, Kansas, 1965
[2] *Glazing Requirements and Procedures for Double Glazing Units*, Insulation Glazing Association, London, 1968

DEFINITIONS OF TERMS CONNECTED WITH FLAT GLASS PRODUCTS

Annealing	Controlled cooling of glass in order to prevent stresses.
Arris edge	A ground or polished bevel on a glass edge.
Bait	The tool dipped into the glass when starting to draw the sheet glass.
Batch	Weighed and mixed raw materials to be fed into the glass furnace.
Bead	A putty or a sealant compound after it has been applied in a joint.
Bevelling	Edge-working process on flat glass.
Bicheroux process	An intermittent rolling process. The glass is cast between rolls on a moving table.
Blister	A relatively large 'bubble' or gas-filled cavity in glass.
Block	A small piece of wood, lead or neoprene used in glazing to position the glass in the frame.
Bloom	A surface film on the glass caused by attack from the atmosphere or vapours.
Blow-pipe	The steel pipe used when blowing glass articles by mouth.
Bruise	A small surface crack caused by impact.
Bullion (*or Bull's eye*)	The central part of the disk made by the crown method.
Cased glass	A glass product which has a surface layer of different colour or composition from that of the main glass body.
Casting	A process of making cast glass usually by pouring the glass onto a table.
Cathedral glass	A rolled glass with a texture on one surface.
Channel	The *U*-shaped opening in a sash or a frame in which the glass panel is placed.

Chemcor glass	A chemically-strengthened glass.
Chemically-strengthened glass	A glass with its surface in compression due to an ion exchange process.
Chip	Breakage of a small fragment out of an otherwise regular surface.
Cladding glass	Usually a toughened glass with a coloured ceramic coating on one surface. Used for curtain walls.
Cord	A glassy inclusion of a composition different from that of the surrounding glass.
Crown glass	A glass made by blowing and spinning—the crown glass method.
Cut size	A glass of cut size is cut from the drawn sheet to a specific dimension.
Cutter	(1) The man who cuts glass. (2) The tool used when cutting glass.
Debiteuse	A clay floating block used in the Fourcault process. The glass is drawn vertically upwards through a slot in the block.
Devitrification	Crystallization in glass.
Dice	The fragments formed from a broken, tempered glass.
Distance piece	A glazing block of wood, lead, neoprene or other suitable material to position the glass in the frame.
Distortion	An optical effect in the glass due to variation of thickness or refractive index.
Draw bar	A clay block used in the Pittsburgh process. The block is submerged in the glass at the point of drawing.
Drawn glass	Sheet glass manufactured in a continuous drawing machine by a vertical or horizontal drawing process.
Etching	Treatment of glass with hydrofluoric acid or other agent for polishing or marking.
Fire-finishing	Surface polishing of glass in a flame.
Fixed light	A window which is not made to be open (in contrary to opening light).
Flashed glass	A coating of glass of different colour or composition to that of the main glass body.

Float glass	A glass made by floating hot glass on a molten metal (the float process).
Fourcault method	A vertical drawing process of making sheet glass by using a slotted debiteuse floater.
Front putty	The putty forming a fillet between the front edge of the rebate and the surface of the glass.
Frosted surface	A surface treatment which scatters the light.
Georgian wired glass	A wired glass with a square pattern of the wire mesh.
Glazing	The setting of glass in prepared openings in windows, door panels, etc. From the interior (inside glazing) or the exterior (outside glazing).
Glazing size	A pane of glass of the correct dimension for glazing.
Grey glass	A heat-absorbing glass.
Heat-absorbing glass	A coloured glass which absorbs a substantial part of infra-red radiation.
Heat-reflecting glass	A coated glass which reflects a substantial part of infra-red radiation.
Heat-resisting glass	A glass which can withstand a high thermal shock. The glass usually contains a high proportion of silica and boron oxide.
Interlayer	The plastic material used in laminated safety glass.
Jamb	The vertical sides of a window or other openings.
Laminated glass	A safety glass of a sandwich construction with a plastic film between the panes of glass.
Leaded light	A decorative window formed by placing small pieces of glass together by means of lead cames.
Lehr	A tunnel-shaped oven for annealing glass.
Libbey-Owens process	Horizontal drawing process for making sheet glass.
Light	Another term for a pane of glass used in windows or other openings.
Metallizing	The process of depositing a metal film on glass.
Obsidian	A glassy material formed by nature.

143

Onion	The mass of glass formed on the top of the debiteuse when drawing flat glass by the Fourcault method.
Opal glass	Glass with a usually white appearance, having light-scattering properties due to tiny inclusions.
Ophthalmic glass	Glass with specific optical qualities used in spectacles.
Pane	(1) A cut size of glass. (2) A division of a sash.
Pennvernon process	A vertical drawing method for making sheet glass using a submerged draw bar.
Photochromic glass	A glass that changes colour according to the intensity of the light that falls upon it.
Pittsburgh process	A vertical drawing method for making sheet glass using a submerged draw bar.
Plate glass	A glass (usually rolled) which has been ground and polished on both surfaces to render them flat and parallel.
Polariscope	An instrument by which the state of annealing in the glass can be controlled by polarized light.
Rabbet	See *rebate*.
Ream	An imperfection in glass caused by a non-homogeneous inclusion.
Rebate	A recess in the sash or frame into which the edge of the glass pane is received (rabbet).
Rolled glass	Flat glass formed by a rolling process.
Roller mark	An imperfection in the surface of sheet glass due to contact with the rollers of the drawing machine.
Rouge pits	Traces of rouge remaining in small cavities on the surface of an incompletely polished glass.
Rough cast glass	Flat glass formed by a casting and rolling process.
Safety glass	A strengthened or laminated glass, which gives better mechanical protection than ordinary glass.
Sand blasting	Projecting a jet of sand onto the surface of the

144

	glass to obtain a matted or obscured surface.
Sash	The frame, including the rebates, to receive the pane of glass in a window.
Seed	A small bubble or gaseous inclusion in glass.
Setting block	See *block*.
Sheet glass	A flat glass made by a drawing process.
Sight size	That part of a window opening which admits light.
Silvering	The process of depositing a film of silver on the surface of glass (when making mirrors).
Solarization	A change in the colour of a glass due to exposure to sunlight.
Sprig	A small headless nail or a thin triangular metal piece used for securing panes of glass in a frame when glazing.
Stained glass	(1) A glass which has a fused, coloured coating on its surface. (2) A glass corroded on its surface by chemical agents.
Stone	A crystalline inclusion in glass.
Striae	Very fine cords.
Substance	The thickness of flat glass.
Tempered glass	A thermally-strengthened glass.
Tong marks	Small circular depressions, near an edge of a piece of toughened glass, caused by the tongs by which it was suspended during tempering.
Toughened glass	A glass which has been mechanically strengthened by a thermal or chemical process in which a layer of compression has been introduced at the surface.
Translucent glass	A glass which transmits light with some diffusion; vision through this glass is not clear.
Wave	Uneven glass distribution in the glass.
Weathering	Chemical corrosion of glass due to attack by atmospheric agents.
Wired glass	A rolled glass containing a wire mesh.

APPENDIX I

UNITS AND CONVERSION TABLES

LENGTH

m	inch	foot	yard
1	39·37	3·28	1·094
0·0254	1	0·083	0·028
0·3048	12	1	0·333
0·9144	36	3	1

1 mile = 1609 m

AREA

m^2	$inch^2$	$foot^2$	$yard^2$
1	1550	10·76	1·196
0·00065	1	0·0069	0·00077
0·0929	144	1	0·111
0·836	1296	9	1

1 acre = 4047 m^2

VOLUME

m^3	$inch^3$	$foot^3$	UK gallon	US gallon
1	61024	35·31	219·97	264·17
$1·639 \times 10^{-5}$	1	0·00058	0·0036	0·0043
0·0283	1728	1	6·228	7·481
0·00455	277·4	0·1605	1	1·801
0·00379	231	0·1337	0·8327	1

MASS

kg	lb	oz	cwt	ton Br
1	2·204	35·27	0·0197	0·00098
0·4536	1	16	0·00893	$4·46 \times 10^{-4}$
0·028	0·063	1	0·00056	$2·79 \times 10^{-5}$
50·80	112	1792	1	0·050
1016·0	2240	35840	0·8929	1

A short ton = 2000 lb   A metric ton = 1000 kg

146

HEAT

Btu	hph	J	kcal	kW h
1	3.929×10^{-4}	1055	0.2520	2.93×10^{-4}
2545	1	2.685×10^{6}	641.3	0.7457
9.48×10^{-4}	3.725×10^{-7}	1	2.389×10^{-4}	2.778×10^{-7}
3.969	1.559×10^{-3}	4187	1	1.163×10^{-3}
3413	1.341	3.60×10^{6}	860	1

$$1 \text{ Btu/ft}^2 \text{ h deg F} = 4.882 \text{ kcal/m}^2 \text{ h deg C}$$
$$4.882 \text{ ft}^2 \text{ h deg F/Btu} = 1 \text{ m}^2 \text{ h deg C/kcal}$$
$$1 \text{ Btu/lb deg F} = 0.309 \text{ kcal/kg deg C}$$
$$1 \text{ kcal/m h deg C} = 8.09 \text{ Btu in/ft}^2 \text{ h degF}$$
$$1 \text{ J} = \text{watt second} = \text{newton metre}$$

LIGHT

Luminous intensity:
  1 candle (International) = 0.981 candela (cd)
Illumination:
  1 foot candle = 10.76 lux (lx)
  $1 \text{ lumen/ft}^2 = 10.76 \text{ lux}$
Luminance:
  1 foot lambert = 3.426 cd/m^2
  $1 \text{ cd/in}^2 = 1550 \text{ cd/m}^2$

VELOCITY

  1 ft/s = 0.3048 m/s
  1 ft/min = 0.00508 m/s
  1 mile/h = 0.447 m/s

PRESSURE AND STRESS

kgf/cm^2 ($=$kp/cm^2)	atm	lb/in^2	lb/ft^2	Newton/m^2
1	0.968	14.22	2050	98070
1.033	1	14.70	2119	101300
0.070	0.068	1	144	6895
4.88×10^{-4}	4.75×10^{-4}	6.95×10^{-3}	1	47.88
1.02×10^{-5}	9.93×10^{-6}	1.45×10^{-4}	2.09×10^{-2}	1

DENSITY

g/cm^3	lb/in^3	lb/ft^3
1	0·0361	62·43
27·68	1	1728
0·016	0·00058	1

TEMPERATURE

$$1 \text{ deg C} = \tfrac{5}{9} (\text{deg F} - 32)$$
$$1 \text{ deg F} = 32 + \tfrac{9}{5} \text{ deg C}$$

deg C	deg F	deg C	deg F
0	32	100	212
5	41	150	302
10	50	200	392
15	59	250	482
20	68	300	572
25	77	350	662
30	86	400	752
35	95	450	842
40	104	500	932
45	113	750	1382
50	122	1000	1832

MULTIPLES AND SUB-MULTIPLES

Symbol	Prefix	Multiplication factor
T	tera	10^{12}
G	giga	10^{9}
M	mega	10^{6}
k	kilo	10^{3}
h	hecto	10^{2}
da	deca	10^{1}
d	deci	10^{-1}
c	centi	10^{-2}
m	milli	10^{-3}
μ	micro	10^{-6}
n	nano	10^{-9}
p	pico	10^{-12}
f	femto	10^{-15}
a	atto	10^{-18}

APPENDIX II

SOME STANDARD SPECIFICATIONS FOR FLAT GLASS

AUSTRIA

Österreichischer Normenausschuss
ÖNORM B 3710. Flachglas Sorten, Dicken Prüfung, Massangabe.

CANADA

Canadian Government Specifications Board
12-GP-1A. Specification for glass: safety.
12-GP-2. Specification for glass: flat, clear sheet.

DENMARK

Dansk Standardiseringsråd
DS 1045. Double pane sealed units.

FINLAND

Suomen Arkkitehtitiitto Standardisoimislaitos
RT 384.11. Koneellisesti vedetty lasi. Laadunmääräykset 1965.

FRANCE

L'Association Francaise de Normalisation
NF B 32-500. Verre. Vitre de sécurité. Terminologie. Classification.
Epaisseurs.
NF B 32-501. Généralités et éprouvettes.
NF B 32-502. Essai de planitude.
NF B 32-503. Essai de résistance des vitres feuilletées à l'immersion
dans l'eau courante.
NF B 32-504. Essai de résistance aux chocs, à la bille, sur éprouvette.
NF B 32-505. Essai de résistance aux chocs, à la bille, sur objet fini.

NF B 32-506. Essai de rupture—contrôle de la limite de résistance aux chocs.

NF B 32-507. Essai de rupture à température normale.

NF B 32-508. Essai de rupture après séjour aux températures extrêmes.

NF B 32-509. Essai de rupture aux températures extrêmes.

NF B 32-510. Essai de rupture au poinçon.

NF B 32-511. Essai de rupture par flexion.

NF B 32-512. Essai de rupture par torsion.

NF B 32-513. Essai de résistance à l'incendie.

NF B 32-514. Essai de déformation de vision.

NF B 32-515. Examen de l'aspect.

NF B 32-516. Essai de résistance aux rayons ultra-violet artificiels.

NF B 32-517. Essai de résistance aux rayons solaires.

NF B 32-518. Essai d'adhérence des feuillets, après exposition aux rayons ultra-violets ou aux rayons solaires.

NF B 32-519. Contrôle de l'influence de l'humidité.

NF B 32-520. Essai de résistance à l'ébullition.

NF B 32-521. Essai de rupture aux chocs à la bille sur vitres armées.

NF B 32-522. Essai au choc d'un sac de lest.

GERMANY

Das Deutsche Normenausschuss

DIN 1249. Tagelglas. Dicken, Sorten Prüfung, Massangaben.

DIN 52301. Sicherheitsglas für Augenschutzgläser. Prüfverfahren.

DIN 52303. Prüfung von Glas. Biegeversuch an Sicherheitsglas.

DIN 52305. Prüfung von Glas. Optische Prüfung von Sicherheitsglas. Bestimmung des Ablenkungswinkels und des Brechtwertes.

DIN 52306. Prüfung von Glas. Kugelfallversuch an Sicherheitsglas.

DIN 52307. Prüfung von Glas. Pfeilfallversuch an Sicherheitsglas.

GREAT BRITAIN

British Standards Institution

BS 952: 1964. Classification of glass for glazing and terminology for work on glass.

BS 3447: 1962. Glossary of terms used in the glass industry.

BS 3193: 1960. Toughened glass doors and panels for domestic appliances and similar uses.

BS 857: 1954. Safety glass for land transport.

JAPAN

Japanese Industrial Standard

J 1S R 3201. Ordinary sheet glass

NETHERLANDS

Nederlands Normalisatie-Instituut

N   1302. Vlakglas. Dikten van de meest voorkomende soorten.

N   3060. Gelaagd veiligheidsglas voor wegverkeersmiddelen. Keuring.

NEN 3264. Spiegelglas. Keuringseisen.

NEN 3265. Vensterglas. Keuringseisen.

NORWAY

Norges Standardiserings-Forbund

NS 1470. Svingvindue, vertikalhengslet med tvillingrute. Type SVT.

NS 1469. Svingvindue, horisontalhengslet med tvillingrute. Type SHT.

NS 1457. Tvillingruter for svingvinduer etter NS 1469 og NS 1470. Formaler og kvaliteter.

SWEDEN

Sveriges Standardiserings-Kommission

SIS 224402. Cast glass.

SIS 224403. Sheet glass.

SIS 224404. Blown glass.

SIS 224405. Plate glass.

SIS 224406. Wired glass.

SIS 224407. Float glass.

SIS 818112. Windows. Types and sizes, fittings and execution.

UNITED STATES OF AMERICA

The Administrator, General Services Administration
DD-G-451a. Federal specification glass, flat and corrugated, for glazing, mirrors and other uses.

American Standards Association, Inc.
Z26.1-1950. American standard safety code for safety glazing materials for glazing motor vehicles operating on land highways.

Department of Housing and Urban Development, Federal Housing Administration
Sections on Flat Glass.
Basic Building Code, Building Officials Conference of America, Inc.
Sections on Flat Glass.

APPENDIX III

SECTIONS FROM THE PROPERTY STANDARDS FOR GLASS

Issued by the Federal Housing Administration, Department of Housing and Urban Development, Washington D.C. and sections dealing with glass from the Basic Building Code, Building Officials Conference of America, Inc., Chicago, Illinois.

711 GLASS

711.1 Standards

711–1.1 All glass for windows and doors, including storm windows and doors, shall comply with F.S. DG–G–451a. Glass shall not be less than 'B' quality.

711–1.2 The glass in all mirrors, including medicine cabinets, shall be plate glass. Mirror construction shall comply with Commercial Standard CS–27.

711–1.3 Glass installed in jalousies shall not be less than $\frac{7}{32}$ inch in thickness nor longer than 36 in. When edges are exposed they shall be seamed or fire-polished.

711.2 Labelling

711–2.1 Each installed pane in windows and doors shall be appropriately labelled. The label shall show the name of the glass manufacturer and the quality and nominal thickness of the glass.

711–2.2 Insulating glass shall be similarly labelled with the fabricator's name.

711–2.3 Tempered or laminated glass shall be permanently labelled in the lower corner. The label shall be visible when installed and show the manufacturer's (temperer's or laminator's) name, type of glass and nominal thickness.

711.3 Maximum areas

711–3.1 Maximum area of glass pane in windows and doors, including

L

storm windows and doors, shall comply with *Table 7–10*, except as modified by 711–3.2, 711–3.3 and 711–4.1. No glass less than 'single strength' thickness shall be used.

711–3.2 When double-glazed units (insulating glass) are used in windows, the maximum area of the pane shall not exceed *Table 7–10* by more than 50% for a thickness corresponding to an individual pane in the same wind zone. Insulating glass shall be a manufactured, hermetically-sealed unit. Both sheets of glass shall be of the same thickness and the air space shall be not greater than $\frac{5}{8}$ in.

711–3.3 For exterior doors, where safety glass is required, the minimum thickness and the maximum area, for equivalent annealed glass in the same wind zone, shall not exceed:

(a) Tempered glass, $\frac{3}{16}$ in minimum; 50% increase over *Table 7–10*.

(b) Laminated glass, $\frac{1}{4}$ in minimum, 60% of *Table 7–10*.

(c) Wire glass, $\frac{1}{4}$ in minimum, 50% of *Table 7–10*.

711.4 Doors

711–4.1 The area, thickness and type of glass used in exterior prime or storm doors, including sliding glass doors, shall be determined as follows:

(a) Single-strength glass may be used when: (1) area of pane is not more than 6 ft², (2) glass is not less than 10 in above the floor, and (3) short dimension of each pane is not more than 15 in.

(b) Double strength glass shall be used when: (1) area of pane is not more than 6 ft², (2) glass is not less than 18 in above the floor, and (3) horizontal muntins or exterior and interior bars supported on stiles are located 34 in (±4 in) above the floor.

(c) At least $\frac{3}{16}$ in annealed glass shall be used when: (1) area of pane is more than 6 ft², (2) glass is not less than 18 in above the floor, and (3) horizontal muntins or exterior and interior bars supported on stiles are located 34 in (±4 in) above the floor.

(d) At least $\frac{7}{32}$ in annealed glass shall be used when: (1) area of pane is more than 6 ft², (2) glass is less than 18 in above the floor, and (3) horizontal muntins or exterior and interior bars supported on stiles are located 34 in (±4 in) above the floor.

(e) Safety glass shall be used when: (1) area of pane is more than 6 ft², (2) glass is less than 18 in above floor, and (3) horizontal muntins or bars are omitted.

(f) At least $\frac{3}{16}$ in annealed glass shall be used in fixed panel of sliding door assemblies.

711–4.2 Where safety glass is required, it shall meet the following tests:

(a) *Fully-tempered glass*

(1) *Particle test*. Tests should be made on a representative number of the largest pieces of glass that will be marketed for use in doors and tub enclosures. The fully-tempered safety glass panel shall be fractured by impact with a spring-loaded centre punch or by hitting a regular centre punch with a hammer. The point of impact shall be $\frac{1}{2}$ in to 1 in from any glass edge. When fractured, there shall be no individual fragment larger than 0·15 oz.

(2) *Impact test*. As in Test No. 8 of ASA Z26.1–1950.

(b) *Laminated glass*

(1) *Boil test*. As in Test No. 4 of ASA Z26.1–1950.

(2) *Impact tests*. As in Tests No. 9 and 12 of ASA Z26.1–1950.

(c) Wire glass, impact test. As in Test No. 11 of ASA Z26.1–1950.

(d) Test data indicating compliance with the above may be required by the FHA field office.

711.5 Shower or tub enclosures

When glass doors or walls are installed in shower stalls or tub enclosures, they shall be safety glass as called for in 711–4.2. Minimum thickness, same as safety glass for doors except that when wire glass is used the thickness shall be not less than $\frac{7}{32}$ in and the area shall not exceed 28 ft^2.

Table 7–10

Wind zone[3]	Maximum glass area[1,2]							
	Nominal glass thickness							
	S.S.	D.S.	$\frac{3}{16}$ in	$\frac{13}{64}$ in	$\frac{7}{32}$ in	$\frac{1}{4}$ in	$\frac{5}{16}$ in	$\frac{3}{8}$ in
	(ft^2)	(ft^2)	(ft^2)	(ft^2)	(ft^2)	(ft^2)	(ft^2)	(ft^2)
Low	10·7	19·5	40·0	48·0	60·0	75·0	90·0	120·0
Medium	7·3	13·2	27·0	32·0	41·0	51·0	62·0	79·0
High	4·8	8·7	18·0	21·0	27·0	34·0	41·0	52·0

[1] Areas in *Table 7–10* apply to regular plate or sheet glass only, and do not apply to special types of glass.

[2] Areas are calculated on a width to length ratio of 1 : 2 or less. Maximum glass areas may be increased in accordance with the following:

From 1 : 2 to 1 : 3 ratio—add 20% to area in *Table 7–10*
From 1 : 3 to 1 : 4 ratio—add 50% to area in *Table 7–10*
From 1 : 4 to 1 : 5 ratio—add 100% to area in *Table 7–10*

[3] See map (*Figure 7a*) for identification of applicable zone.

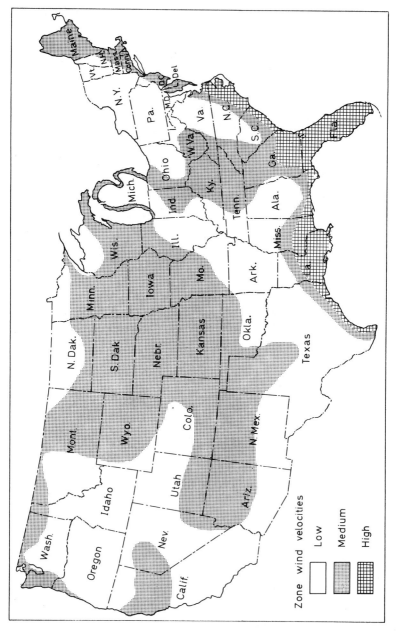

Figure 7a

Sec. 714.4 (S18-63-64 Pt. 1)

714.4 *Design wind load for glass*—Firmly-supported lights of glass of 4 ft² or more in area, installed in a vertical position, or at an angle of not more than 20° from the vertical, shall be designed to withstand wind pressure in accordance with design criteria stated in Appendix K-12.

Sec. 803.1 (S19-63-64)

803.11 *Strength tests for glass*—The working strength of glass for any location in which it is required to withstand specific loads shall be determined as provided in Appendix K-12-B.

Sec. 859.5 (S20-63-64)

859.5 Glass in walls

859.51 *Labelling*—Each light of glass shall be labelled with a removable paper label showing type, thickness and manufacturer. To qualify as glass with special performance characteristics, each unit of laminated, heat-strengthened, fully-tempered, and insulating glass shall be permanently identified by the manufacturer. The identification shall be etched or ceramic-fired on the glass and be visible when the unit is glazed.

Heat-strengthened and tempered spandrel glasses are exempted from permanent labelling. This type of glass shall be labelled with a removable paper label by the manufacturer.

859.52 *Glass dimensional tolerance*—Glass thickness tolerances shall comply with those established in the following table (see also p. 158). Where thickness is to be controlled, nominal values are stated subject to these tolerances:

Nominal thickness	Plate glass min. thickness	Sheet glass min. thickness
	(in)	(in)
S.S.	—	0·085
D.S.	—	0·115
$\frac{1}{8}$	0·094	—
$\frac{3}{16}$	0·156	0·182
$\frac{13}{64}$	0·172	—

157

Nominal thickness	Plate glass min. thickness	Sheet glass min. thickness
$\frac{7}{32}$	—	0·205
$\frac{1}{4}$	0·218	0·240
$\frac{5}{16}$	0·281	—
$\frac{3}{8}$	0·312	0·312
$\frac{1}{2}$	0·437	0·438
$\frac{5}{8}$	0·532	—
$\frac{3}{4}$	0·625	—
1	0·875	—
$1\frac{1}{4}$	1·125	—

859.53 *Glass supports*—Where one or more sides of any light of glass is not firmly supported, or is subjected to unusual load conditions, detailed shop drawings, specifications and analysis by methods described in Appendix K, or test data assuring safe performance for the specific installation, shall be prepared by engineers experienced in this work and approved by the building official.

859.54 *Wind loads*—Glass exposed to wind pressure shall be capable of safely withstanding the loads specified in *Table A* in Appendix K, acting inwards or outwards. The safety factor relating the maximum working stress to breaking stress shall be not less than two and a half. Owners or tenants shall replace cracked lights promptly.

Glass which conforms to the required nominal thickness of *Table C* in Appendix K for the design wind loads of *Table A* in Appendix K shall be accepted.

859.55 *Jalousies*—In jalousie windows and doors regular plate, sheet or rolled glass thickness shall be not less than $\frac{3}{16}$ in; glass length shall be not more than 36 in; glass edges shall be smooth. Other types of glass may be used if detailed shop drawings, specifications and analysis by methods described in Appendix K, or test data assuring safe performance for the specific installation, are prepared by engineers experienced in this work and approved by the building official.

859.56 *Human impact loads*—Glass in prime and storm doors, interior doors, fixed glass panels which may be mistaken for means of egress or ingress, shower doors and tub enclosures shall meet the requirements set forth in the following table, or by comparative tests shall be proven to produce equivalent performance.

The following shall be used as criteria for determining the safety hazard:

(a) Glass in openings of regularly occupied spaces.

(b) Lowest point less than 18 in above finished floor.

(c) Minimum dimension larger than 18 in.

IMPACT LOADS—GLASS

Glass shall conform to these limits:

(1) Glass less than single strength (S.S.) in thickness shall not be used.

(2) If short dimension is larger than 24 in, glass must be double strength (D.S.) or thicker.

Glass type	Individual opening area (ft²)	Requirements
Regular plate, sheet or rolled (annealed)	Over 6	Not less than $\frac{3}{16}$ in thick. Must be protected by a pushbar or protective grille firmly attached on each exposed side*, if not divided by a muntin
Regular plate, sheet or rolled (annealed), surface sandblasted, etched, or otherwise depreciated	Over 6	Not less than $\frac{7}{32}$ in thick. Must be protected by a pushbar or protective grille firmly attached on each exposed side*
Regular plate, sheet or rolled (annealed), obscure	Over 6	Not less than $\frac{3}{16}$ in thick. Must be protected by a pushbar or grille firmly attached on each exposed side*
Laminated	Over 6	Not less than $\frac{1}{4}$ in thick. Shall pass impact test requirements of Appendix K-12
Fully-tempered	Over 6	Shall pass impact test requirements of Appendix K-12
Wired	Over 6	Not less than $\frac{1}{4}$ in thick. Shall pass impact test requirements of Appendix K-12
All unframed glass doors (swinging)		Shall be fully-tempered glass and pass impact test requirements of Appendix K-12

* Building owners and tenants shall maintain push-bars or protective grilles in safe condition at all times.

APPENDIX B

ACCEPTED ENGINEERING PRACTICE STANDARDS

(S18-63-64 Pt. 3) (S27-63-64) (S13-64)

(S25-64 Pt. 1) (S35-64 Pt. 2) (S38-64 Pt. 3).

Glass

Safety glazing materials for glazing motor vehicles
   operating on land highways, safety code for......ASA Z26.1—1950
   Fully-tempered ....No. 8, ASA Z26.1—1950
   Laminated ....Nos. 4, 9 and 12, ASA Z26.1—1950
   Wired ....No. 11, ASA Z26.1—1950

APPENDIX K

UNIT WORKING STRESSES FOR ORDINARY MATERIALS
(S18-63-64 Pt. 2)

Add new section and subsections as follows:

K-12 GLASS DESIGN CRITERIA

When glass is used as a major portion of walls of buildings and is required by the code to withstand wind or impact loads, or to meet minimum thickness standards, the following design procedure and criteria shall be used.

K-12-A Design wind loads for glass

The wind pressure in pounds per square foot to be used in the design of glass as required in Section 714.4 shall be that shown in *Table A* corresponding to the opening elevation and wind velocity shown in *Table B* for the geographic location of the building.

K-12-A-1 Elevation and terrain—The elevation of a glazed opening shall be computed by adding the distances from grade to the head and sill, respectively, and dividing the sum by two. Elevation correction factors given in *Table A* shall be applied to the fastest mile wind velocity in miles per hour. Correction factors for terrain given in *Table A* also shall be applied to the fastest mile wind velocity in

Table A. Design wind load—glass in pounds per square foot at various elevations above grade for various fastest mile wind velocities*
(50 year recurrence interval)

Elevation above grade (ft)	Fastest mile velocities, V* Design loads, p for flat open country																					
	V	p	V	p	V	p	V	p	V	p	V	p	V	p	V	p	V	p	V	p	V	p
0–10	42	6	46	7	49	8	52	9	55	10	59	11	62	12	66	14	69	15	76	19	83	22
10–20	52	9	58	11	61	11	65	14	70	16	74	18	79	20	83	22	87	24	96	30	105	35
20–30	60	12	67	14	70	16	75	18	80	20	85	23	90	26	95	29	100	32	110	39	120	46
30–60	66	14	74	18	77	19	83	22	88	25	94	28	99	31	104	35	110	39	121	47	132	56
60–120	73	17	82	21	85	23	92	27	98	31	104	35	110	39	116	43	122	48	134	57	146	68
120–240	81	21	91	26	95	29	101	33	108	37	115	42	122	48	128	52	135	48	149	71	162	84
240–480	90	26	100	32	104	35	112	40	119	45	127	51	134	57	142	65	149	71	164	86	179	102
480–960	98	31	110	39	115	42	123	49	131	55	139	62	148	70	156	78	164	86	180	104	197	124
over 960	98	31	110	39	115	42	123	49	131	55	139	62	148	70	156	78	164	86	180	104	197	124

* Select from U.S. Weather Bureau Map, *Table B.*

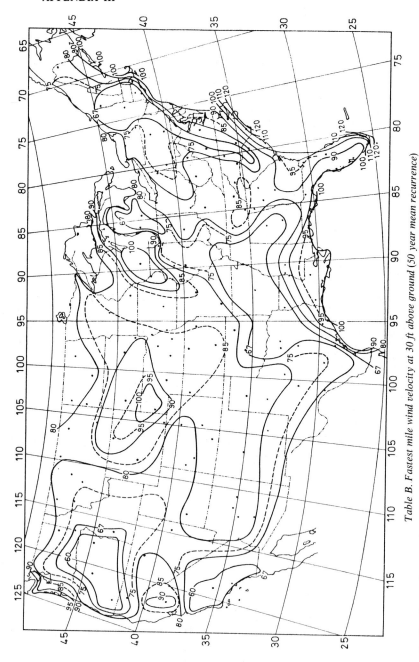

Table B. Fastest mile wind velocity at 30 ft above ground (50 year mean recurrence)

Speeds are for normal exposure where surface friction is relatively uniform for a fetch of about 25 miles. If the exposure is elevated, subject to channelling or other special conditions affecting the extreme wind speeds, adjustments must be made to the map values.

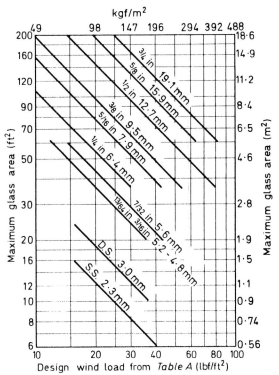

Table C. Required nominal thickness of regular plate or sheet glass

Based on minimum thicknesses before weathering, allowed in Federal Specification DD-G-451a. Design factor = 2·5.

163

miles per hour. These factors shall not be applied to the design wind load in pounds per square foot.

K-12-B Impact loads

To qualify as materials of special impact resistance characteristics, fully-tempered, laminated and wired glass shall comply with the requirements of the applicable standards listed in *Appendix B*.

Table D. Relative resistance to wind load
(assuming equal thickness)

Glass type	Approximate relationship*
Laminated	0·6
Wired glass	0·5
Heat-strengthened	2·0
Fully-tempered	4·0
Factory-fabricated double glazing†	1·5
Rough-rolled plate	1·0
Sand-blasted	0·4
Regular plate or sheet	1·0

* Before using *Table C* divide the design wind load from *Table A* by the value shown here for the glass type involved.

† Use thickness of the thinner of the two lights, not thickness of unit.

INDEX

165

DATE DUE

GAYLORD			PRINTED IN U.S.A.